Urban Growth Patterns in India

Urban Growth Patterns in India

Spatial Analysis for Sustainable Development

Bharath H. Aithal and T. V. Ramachandra

With the collaboration of

Chandan M. C.
Nimish G.
Vinay S.
Aishwarya N.

CRC Press
Taylor & Francis Group
Boca Raton London New York

CRC Press is an imprint of the
Taylor & Francis Group, an **informa** business

CRC Press
Taylor & Francis Group
6000 Broken Sound Parkway NW, Suite 300
Boca Raton, FL 33487-2742

and by CRC Press
4 Park Square, Milton Park, Abingdon, Oxon, OX14 4RN

First issued in paperback 2022

© 2020 by Taylor & Francis Group, LLC
CRC Press is an imprint of Taylor & Francis Group, an Informa business

No claim to original U.S. Government works

ISBN 13: 978-1-03-247472-4 (pbk)
ISBN 13: 978-0-367-22521-6 (hbk)
DOI: 10.1201/9780429275319

Publisher's Note
The publisher has gone to great lengths to ensure the quality of this reprint but points out that some imperfections in the original copies may be apparent.

Visit the Taylor & Francis Web site at
http://www.taylorandfrancis.com

and the CRC Press Web site at
http://www.crcpress.com

Dedicated

to

The Society

&

My Teachers, Parents, Brother, and Friends

—Bharath H. Aithal

Disclaimer

The views expressed in this book are those of the authors and do not necessarily reflect the views or policies of the Indian Institute of Science, Bangalore, India.

Bharath H. Aithal

Contents

Preface xi

Acknowledgements xiii

Authors xv

1 Landscape Dynamics: An Introduction 1
  M. C. Chandan, Bharath H. Aithal and T. V. Ramachandra

2 Quantification of Greenhouse Gas Footprint 19
  G. Nimish, Bharath H. Aithal and T. V. Ramachandra

3 Land Use and Land Cover Dynamics: Synthesis of
  Spatio-Temporal Patterns 39
  M. C. Chandan, Bharath H. Aithal and T. V. Ramachandra

4 Spatial Metrics: Tool for Understanding Spatial Patterns of
  Land Use and Land Cover Dynamics 61
  M. C. Chandan, Bharath H. Aithal and T. V. Ramachandra

5 Land Use Modelling: Future Research, Directions and Planning 81
  M. C. Chandan, Bharath H. Aithal and T. V. Ramachandra

6 Current Trends in Estimation of Land Surface Temperature
  Using Passive Remote Sensing Data 107
  G. Nimish, Bharath H. Aithal and T. V. Ramachandra

7 Sustainable Development Goals (SDGs): Disaster Mitigation in
  Flood-Prone Regions of India 127
  Narendr Aishwarya, Bharath H. Aithal and T. V. Ramachandra

8 Spatial Decision Support System (SDSS) for Urban Planning 149
  T. V. Ramachandra, Vinay Shivamurthy and Bharath H. Aithal

Index 173

Preface

Urbanisation refers to the growth of towns and cities due to a large proportion of the population living in urban areas and its suburbs at the expense of its rural areas. Unplanned urbanisation leads to large-scale land use changes affecting the sustenance of local natural resources. This necessitates an understanding of spatial patterns of urbanisation to implement appropriate mitigation measures. It is estimated that about 54% of the world's population resides in cities and as per the World Urbanization Prospects, the urban population growth poles will be in South East Asia and Africa. A planned, sensible urbanisation process provides adequate infrastructure and basic amenities while maintaining the environmental balance.

Indian cities have been experiencing rapid urbanisation subsequent to globalisation and the opening of Indian markets during the last two decades. Local governments, land use managers and city planners are challenged to meet the basic service needs of the residing population and large fluxing rural–urban population and to provide the basic amenities and infrastructure required in step with growing populations. Each city in India has a varied growth pattern based on its demography, economy, connectivity, location, etc. Cities also provide resilient communities with livelihoods for their citizens; this also entails protection of the local environment and precious natural resources. Changes in the land use pattern need to be understood and modelled for advance geovisualisation of likely growth pockets. Thus, understanding the spatio-temporal patterns of urbanisation in India and its implications on local environs and climates constitute the prime goal of this book. This process entails monitoring urban dynamics to understand the spatial patterns of urbanisation and identifying rapid growth poles. Understanding the behaviour of urban dynamics with the insights of the region's carrying capacity would aid in evolving appropriate strategies towards the design of sustainable cities and estimating resources necessary for the future support of the population. This book aims to understand the spatial patterns of land use/land cover (LULC) dynamics in the rapidly urbanising cities of India and to simulate future urbanisation patterns considering a 10 km buffer using the SLEUTH model. The spatial patterns of landscape dynamics are assessed using temporal remote sensing data of 1992 and 2017 acquired from Landsat repositories. Based on these landscape dynamics, SLEUTH was employed to understand the changing trends for visualising future patterns. Further, the concentration of greenhouse gases (GHG) in the atmosphere has rapidly increased due to anthropogenic activities resulting in a significant increase in the temperature of the Earth due to global warming. This book will also measure environmental sustainability through focus on accounting for the amount of three important greenhouses gases, namely carbon dioxide (CO_2), methane (CH_4) and nitrous oxide (N_2O), and thereby developing a GHG footprint of the major cities in India. Finally, the book will illustrate in detail and analyse urban floods and how spatial temporal analysis can aid in understanding this aspect and act as a tool for developing an indicator for achieving sustainable development. The analyses presented in this book can be used to engage in discussion about methods of sustainable development and policy narratives which are required to build smart, environmentally friendly cities. This book seeks to serve three purposes. First, it can be used as a reference material/publication by researchers as base reference for their next level of research. Second, it can define Sustainable Development Goals (SDGs) through a unique way of understanding a bottom-up approach of cities and cityscape. Third, this book can be used in undergraduate-level courses for case study analysis of enviro-spatial analysis. In all three aspects, the goal is to provide a holistic view of developing cities, based on a sustainable agenda, to a wide audience.

Acknowledgements

We take this opportunity to honour and thank each and every individual who has been instrumental in making this book a reality. We would like to thank all authors and in-house reviewers who have been critical with inspiring words. We also thank proposal reviewers for their encouraging words.

I (B. H. Aithal) also would like to acknowledge the love, help and appreciation that I received from my parents Haridas P. Aithal and Mahalakshmi, brother Ar. Sharath H. Aithal and wife Gouri Bharath Aithal whose consistent support helped in developing the book on time. I would also like to thank my entire family for their support. My love and thanks to my sweet kids Adhira and Aarav for always keeping me energised.

We would also like to thank and follow in the footsteps of Swami Vivekananda and Bhagat Singh, people of great wisdom, teaching and knowledge who were also major sources of inspiration for our teaching.

We are extremely grateful to Prof. Bhargab Maitra, head of Ranbir and Chitra Gupta School of Infrastructure Design and Management, for his support. We also thank Prof. Joy Sen and Mr. Ranbir Gupta of IIT Kharagpur and Prof. N. V. Joshi for their support and encouragement, and their critical and logical comments which inspire us to contribute to the betterment of the science community. We also thank other professors at the Ranbir and Chitra Gupta School of Infrastructure Design and Management for their advice and timely help. Prof. Mukund Dev Behera, Prof. Arkopal Goswami and Prof. Sutapa Das also deserve special mention for their timely help and encouragement.

We are grateful to the Taylor & Francis staff who have been extremely helpful and very involved in developing this book. Our heartfelt thanks to all the members of EWRG (IISc) and EURG (IIT Kharagpur) for supporting and providing their help.

Our sincere thanks to the National Remote Sensing Centre (NRSC), Hyderabad, for providing the IRS data, and the Global Land Cover Facility and National Aeronautics and Space Administration (United States), and the United States Geological Survey for providing the Landsat imageries. We thank ISRO for providing satellite data and visualisation tools through Bhuvan. We thank the Department of Science and Technology (DST) of the Government of India and the Department of Higher Education (West Bengal DST), and the Indian Institute of Technology Sponsored Research cell for all infrastructure and financial assistance.

Authors

Bharath H. Aithal, PhD, is an assistant professor at Ranbir and Chitra Gupta School of Infrastructure Design and Management at the Indian Institute of Technology, Kharagpur, India. He earned his PhD in landscape modelling from the Division of Engineering Sciences, Indian Institute of Science, Bangalore, India. He has published 40 peer-reviewed journal papers, 4 book chapters, 75 conference proceedings, and 42 technical reports during the last 90 months in the subjects of pattern analysis and urban modelling. He has been invited to deliver 18 guest lectures at various Indian academic institutions, universities, and research and development organisations. His works on the effects of thermal plants on land use change and the environment in Udupi and Yettinahole were major contributions in Udupi and Mangalore (all in India). His works on the urbanisation of various cities of India has been appreciated and adopted by city planners, experts and political visionaries. He specialises in urban pattern analysis (spatial), data modelling, machine learning and enviro-socio analysis.

T. V. Ramachandra, PhD, is a coordinator of the Energy and Wetlands Research Group (EWRG) and convener of the Environmental Information System (ENVIS) at the Centre for Ecological Sciences (CES), Indian Institute of Science (IISc). He earned his doctoral degree in energy and environment from the Indian Institute of Science. He has made significant contributions in the area of energy and environment. His research areas include wetlands, conservation, restoration and management of ecosystems, environmental management, GIS, remote sensing, regional planning and decision support systems. He teaches principles of remote sensing, digital image processing and natural resources management. He has published more than 250 research papers in reputed peer-reviewed national and international journals and 200 papers for national and international symposiums, as well as 14 books. In addition, he has delivered a number of plenary lectures at national and international conferences. He is a fellow of the Institution of Engineers (India) and IEE (UK); and a senior member of the IEEE (USA) and many similar institutions. Details of his research and copies of publications are available at http://ces.iisc.ac.in/energy/ and http://ces.iisc.ac.in/grass.

1

Landscape Dynamics

An Introduction

CONTENTS

1.1 Introduction 1
1.2 Geospatial Approaches in Assessing Landscape Dynamics 2
1.3 Spatial and Temporal Pattern Assessment of Landscapes 4
1.4 Land Use Assessment 6
1.5 Landscape Structure Measurement and Indices 8
1.6 Land Use Modelling 9
1.7 Indian Context 12
1.8 Concluding Remarks 15
References 15

1.1 Introduction

Landscape is a mosaic of interacting dynamic ecosystems. The composition of various landscape elements determines the structure of a landscape, which determines the functioning of respective ecosystems. The structure of a landscape, or natural features in a landscape, is influenced either by natural forces or altered by anthropogenic activities. Alterations in the landscape structure vary in terms of size, shape, configuration and arrangement (Antrop, 1998; Turner, 1987). Changes in land use and land cover alter the structure of a landscape and impact the local ecosystem and the global environment (Foley et al., 2005) with gradual or sudden changes in the functional capability of an ecosystem, which affects human lives. Landscape provides a vital link between living organisms and the surrounding environment. Fragmentation of landscape involves the conversion of a single patch into multiple patch types, for instance, conversion of a forest area into a small human settlement followed by agriculture and other land use forms (Saunders et al., 1991; Forman, 1995). Sudhira et al. (2003) outlined the evolution of human settlements from villages to huge megacities. These cities, sometimes also referred to as urban centres, are the result of natural landscapes losing their identity with complex fragmentations. Understanding land use/land cover

(LULC) dynamics in a region is essential to assess the level of fragmentation. Land cover (LC) can be defined as the types of physical features present on the earth surface (Lillesand et al., 2012). LC reflects the visible evidence of existing cover pertaining to vegetation and non-vegetation or water and non-water (Ramachandra et al., 2013) on the surface. Land use (LU) is generally referred to as socio-economic utilisation of land due to human activities. It includes exploited land cover types used for industrial, agricultural, plantation, commercial and residential purposes. The alteration of the landscape structure through large-scale land use/land cover change (LULCC) would impact the nutrient, water and biogeochemical cycles directly or indirectly (Ramachandra et al., 2012; Paudel & Yuan, 2012). LULCC is related to anthropogenic activities and affects the biophysics, biogeochemistry and biogeography of the earth's surface and atmosphere. The effects may be in terms of long wave radiations and violent atmospheric conditions such as changes in heat, water vapour, temperature, carbon dioxide and other trace gases (Kabat et al., 2004; NRC, 2005; Cotton & Pielke, 2007). This necessitates understanding LULCC to formulate appropriate mitigation strategies. Spatial data acquired through space-borne sensors with recent advances in geo-informatics form an overhead perspective at regular intervals (remote sensing data), which help in mapping LULC dynamics.

1.2 Geospatial Approaches in Assessing Landscape Dynamics

Geographic information system (GIS) and remote sensing techniques are useful in assessing spatial and temporal changes in the landscape, which are essential in regional planning for evolving sustainable management of natural resources to achieve sustainable development (Sudhira et al., 2003). Multi-resolution spatial data of remote sensing has been useful in deciphering spatial and spectral heterogeneity of urban environments (Jensen & Cowen, 1999) and is emerging as one of the best available technique to analyse wider spatial extent. Remote sensing refers to "the science and art of obtaining information about an object, area, or phenomenon through the analysis of data acquired by a device that is not in contact with the object, area or phenomenon under investigation" (Lillesand et al., 2012). The spatial information acquired at regular intervals through sensors mounted in the satellites aids in the inventorying, mapping and monitoring of Earth's resources. The remote sensing process is governed by seven general steps as indicated in Figure 1.1.

GIS is useful to capture, store, query, analyse and display geospatial data (Chang, 2015). GIS has two major data types: geographic (spatial) and attribute. Spatial data deals with all aspects of location, geometry and shape of the object or area, whereas the attribute data implicates characteristics

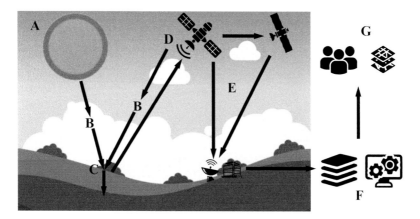

FIGURE 1.1
Remote sensing process. (A) Energy source, the sun (passive remote sensing) or through another source that can transmit its own energy (active remote sensing). (B) Transmission of energy through Earth's atmosphere. (C) Energy interactions at the earth surface. (D) Energy emitted by the earth surface is recorded by sensors. (E) Generation of raw data (essentially in pictorial or digital format) and communication between satellites and ground stations. (F) Processing, analysis and interpretation of data with the help of computer systems. (G) With the help of GIS, various layers are taken in an ordered fashion to visualise and produce useful data to users.

associated with each of the spatial features. The GIS data model has two major divisions. The first is the vector data model (Figure 1.2), which uses spatial (X, Y) coordinate values (also referred as location coordinates) to construct spatial features (points, lines and areas/polygons). Second, the raster data model uses a collection of grids or pixels (picture elements) to represent the spatial variation of a feature (Figure 1.2).

GIS functions include handling and managing spatial and attribute data, displaying and designing maps, data exploration and editing as well as modelling. The ability of GIS to handle and process geospatial data makes GIS distinct from other information systems. GIS has a wide range of applications such as natural resources management, climate change, urban planning, transportation routing, emergency planning, land use change detection and crime analysis. Much of the geoinformatics capability of GIS comes from the database management system (DBMS), which is designed to store and retrieve any kind of attribute data. DBMS involves commands and query languages, for instance SQL. There are various proprietary and open source software programmes available in the market for GIS.

The Global Positioning System (GPS) helps in ascertaining the spatial coordinates (X, Y) of a location, which have immense value in navigation, spatial analyses, etc. GIS and GPS are more or less interdependent on each other. Every point on the earth surface has its own coordinate or XY values. These values can be recorded by a minimum of four communication satellites to give precise location data. The satellites operate at an approximate height of 20,000

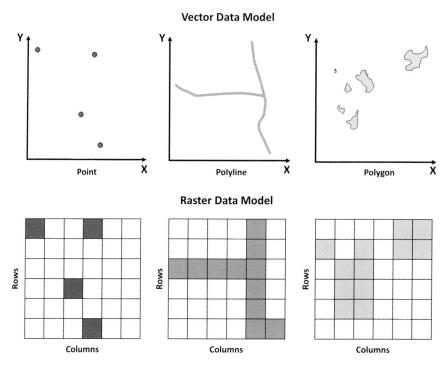

FIGURE 1.2
Vector and raster data models.

kilometres from the earth surface. Many nations have their own positioning systems. For example, Russia has its Global Navigation Satellite System (GLONASS), and most recently India came up with the Indian Regional Navigation Satellite System (IRNSS) aimed to provide accurate location services for military purposes. GPS helps to improve mapping activities by providing ground truth or reference data. The decision support systems within GIS help in the decision making through geospatial data sets with a multi-agent evaluation (Axtell & Epstein, 1994; Parker et al., 2003) that helps to monitor, model and predict landscape dynamics, land use/land cover changes, etc.

1.3 Spatial and Temporal Pattern Assessment of Landscapes

Changes in landscapes and their implications on humans and other life forms have necessitated large-scale research throughout the globe. Over the last two or three decades, the interest has shifted from identifying or detecting landscape dynamics to modelling future landscape transitions and exploring possible trends of basic level changes. Assessment of landscape dynamics has always been regarded by various disciplines apart from fundamental land-scape ecology. Even though the building concepts and techniques are still attributed to ecology, researchers with remote sensing backgrounds have con-sistently applied monitoring landscape dynamics. There is a growing need for landscape monitoring since city managers in megacities face unprecedented challenges with regard to landscape planning and land use management due to high dynamic growth, changes in spatial patterns over time and failures to identify driving forces and pockets for landscape changes, especially in developing countries like India, China and those in Africa. This necessitates understanding the temporal changes in the land use and rate of growth that would help in providing vital insight to the decision-making process through understanding impacts of landscape changes, biodiversity, complexity and fragmentation of the landscape. Therefore, the quantification of landscape changes considers both modifications in spatial arrangements and their con-sequences. Concepts and advantages of spatial data available temporally pro-vide a diversified approach towards assessing landscape structure on both spatial as well as temporal scales (Houet et al., 2009). Spatial resolution of remote sensing refers to the ability to distinguish between two closely spaced objects on an image. It gives a perspective of coarseness or fineness of a raster image. Spatial resolution is expressed by the size of the pixel representing the ground in meters and controlled by the instantaneous field of view (IFOV) of a sensor or detector. It is expressed as a solid angle through which a detector is sensitive to the radiation. IFOV also depends on the altitude of the satellite above ground level and the viewing angle of the sensor. The finer the spatial resolution, the greater the detail of the ground. For instance, Landsat-8 has a spatial resolution of 30 m, whereas Sentinel-1A has a 5 m resolution. This means a pixel of Landsat can hardly differentiate between a building and any other category within 30 × 30 m, but spatial data acquired from Sentinel can easily differentiate building types.

Temporal resolution is the measure of an object on the ground with respect to time. Temporal resolution for satellites may vary from hours to days. Actual temporal resolution of a sensor depends on the satellite/sensor capa-bilities and swath overlap. For instance, Landsat-8 has a temporal resolution of 16 days. Figure 1.3 shows Landsat-8 imagery for a part of Delhi, the capital city of India, during different time periods. It is to be noted that in this case the temporal resolution is 5 years, since the images were acquired during the years 2008, 2013 and 2018.

Spatial data analyses involve the generation of either a true colour compos-ite (TCC) or false colour composite (FCC) to take advantage of spectral resolu-tions of the acquired spatial data. True colour composite uses the spatial data acquired in blue (0.4–0.5 μm), green (0.5–0.6) and red (0.6–0.7) wavelengths, which are assigned the true colours blue, green and red, respectively, which closely resemble the observation made by the human eye. However, colour assignment of any band of a multi-spectral data can be assigned RGB in an arbitrary manner. This kind of colour assignment apart from the natural

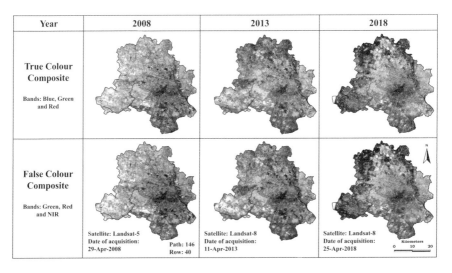

FIGURE 1.3
True and false colour composite images of Delhi during different time periods.

colour composite is known as FCC. There are many possible combinations of FCC depending upon the use and purpose of the spatial data application. For example, spatial data in FCC allows users to detect various features readily due to heterogeneity, for example, healthy vegetation appears to be red and urban areas appear to be bluish-grey in nature in an FCC.

Remotely sensed data acquired through space-borne (satellite platform) sensors are able to capture large areas in an instant allowing users to analyse landscape dynamics with the spatial information obtained on structure and composition at regular intervals. Unlike traditional surveying methods, remote sensing technology helps in acquiring information in a very cost-effective manner and less time, and avoids tedious, laborious work by professionals. However, there are several unaddressed challenges related to resolution trade-offs and cost issues when it comes to very high-resolution spatial data, security issues with digital data and so on. Of late, the use of fine-scale spatial and temporal imageries has proved efficient in monitoring landscape dynamics across the globe and much of the power comes when a user decides to combine multi-scale, multi-sensor and multi-source data to generate highly reliable land use data for monitoring and mapping (Gardner et al., 2008).

1.4 Land Use Assessment

LULCC leading to deforestation has been a prime driver of global climate change. Population growth and the consequential increase in global economic development have led to a significant migration of people to the urban parts of the world. In addition, LULCC has been referred to as the products or outcomes of the prevailing interaction between natural and anthropogenic factors and their utilisation by man in time and space. Performing a detailed analysis of LULCC over an area can help in explaining the spatial extent and the degree of the change itself, and help in assessing the directions and degree of other human-related environmental changes. Assessing land use is one of the basic steps towards achieving sustainable development goals. Traditional field data are limited to a specific extent. They have to be processed and then brought to a common platform to be applied on regional or global extents, whereas land use maps from remotely sensed data have a unique advantage of analysing large areas in very limited time. Classification is one of the most known techniques to derive information from raw satellite data in terms of land categories. Categorisation is completely based on an LULC classification system developed by Anderson (1976). There are two broad categories of classifying spatial data: (a) supervised (based on field ground truthing data) and (b) unsupervised classifiers. Details of types of classifiers are illustrated in Figure 1.4. The former technique needs user intervention to classify data sets based on the training data sets. The latter does not involve the user; however, it classifies the data based on natural grouping/clustering. Researchers have characterised LU using a knowledge-based supervised classification technique to achieve accurate results. The Gaussian maximum likelihood classifier (GMLC) algorithm is one such technique that is well established, tested and adopted by numerous researchers worldwide. GMLC is the most reliable, consistent and well-regarded technique since it considers not only the mean, variance and covariance but also probability

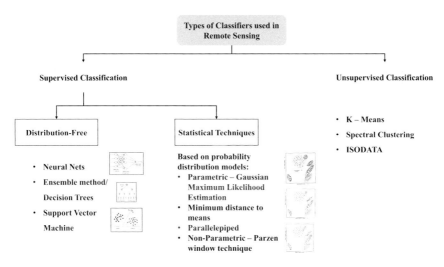

FIGURE 1.4
Types of classifiers used in LULC mapping.

density contours to assign an unknown pixel to a group/category. Land use classification is validated through accuracy assessments in comparison with the reference data. An accuracy assessment can be performed by comparing the classified map with the validation map. The error/confusion matrix generated helps to estimate user's accuracy (error of commission), producer's accuracy (error of omission), overall accuracy and the kappa coefficient (Congalton, 1991).

$$\text{Overall Accuracy} = \frac{\text{Sum of diagonal elements}}{\text{Total number of pixels}}\% \qquad (1.1)$$

$$\text{Kappa} = \frac{\text{Observed accuracy} - \text{chance agreement}}{1 - \text{chance agreement}} \qquad (1.2)$$

1.5 Landscape Structure Measurement and Indices

Assessment of landscape structure entails delineating homogeneous/heterogeneous areas along with the quantification, their spatial arrangement and diversity. Understanding landscape structure has been the primary focus of researchers in the field of landscape ecology and land use managers. Quantification of landscape structures began in the 1980s with the patch matrix model, initially used in the field of landscape ecology, with attempts to identify a symbiotic relationship between landscape structure and the surrounding ecology (Lausch et al., 2015). Landscape metrics have been developed to measure the structural composition of a landscape. The spatial component of landscape analysis can be of benefit if it is considered along with spatial data (acquired at regular intervals through space-borne sensors, that is remote sensing data). There is no single metric that can capture an entire landscape structure or change dynamic, though many of these metrics are associated with one another. Metrics are sensitive to the scale of assessment and they have the ability to determine results statistically. These metrics are further discussed in Chapter 4. Similar to metrics, there are indices which help define the structure of a landscape. Four different indices based on the literature review (Kasanko et al., 2006; Qiuying et al., 2015; Wu et al., 2015)—annual increase index (AII), annual land use growth rate index (ALGRI), annual average rate of change (AARC) and land use expansion intensity index (LEII)—were used to compare and estimate the magnitude of landscape dynamics. These indices help to identify the type of landscape expansion, complexity, fragmentation and disintegration with respect to other landscapes. AII is used to measure annual changes in land use, whereas the ALGRI eliminates the size aspect of a landscape and is suitable for comparison of landscapes at different time periods. AARC helps to

TABLE 1.1

Landscape Indices

Index	Index Type and Formula	Range
AII	$$\text{AII} = \frac{L_{\text{final}} - L_{\text{start}}}{t}$$	AII > 0, without limit
ALGRI	$$\text{ALGRI} = \left[\left(\frac{L_{\text{final}}}{L_{\text{start}}} \right)^{1/t} - 1 \right] * 100\%$$	$0 \leq \text{ALGRI} \leq 100\%$
	AII (km^2 per year) and ALGRI (%), L_{start} and L_{final} refers to land use at starting and end of time period considered for analysis, t being the time span between start and end in years.	
AARC	$$\text{AARC} = \left[\frac{L_{\text{final}} - L_{\text{start}}}{L_{\text{start}}} \right] * \left(\frac{1}{t} \right) * 100\%$$	$0 \leq \text{AARC} \leq 100\%$
LEII	$$\text{LEII} = \left[\frac{L_{\text{final}} - L_{\text{start}}}{t} \right] * 100/\text{TA}$$	LEII > 0, without limit
	TA is the total area. LEII can be classified into five categories: <0.1, very low 0.1–0.2, low 0.2–0.4, moderate 0.4–0.7, rapid ≥0.7, highly rapid	

capture temporal landscape dynamics and changes over time. LEII describes the intensity of landscape change pattern during different time periods. The indices used and their descriptions are given in Table 1.1.

1.6 Land Use Modelling

Understanding landscape dynamics on spatial and temporal scales using a modelling approach is complex due to the factors driving changes and the behaviour within a specified environment. Prevalent usage of multi-resolution remote sensing data (satellite data) with spatial scales ranging from metres to several hundred kilometres has a variety of applications for Earth observation. These complexities have been addressed with the help of various modelling strategies developed in the last two decades such as cellular automata models, artificial intelligence and machine-learning-based models, and agent-based models. These approaches aid in land management and also help in assessing the future of land with various scenarios. Models designed for land use management allow users to test, calibrate, validate and predict near-future trends and scenarios.

Models incorporating statistical data sets, rules of transitions and tools aid in simulating human–environment systems to achieve realistic results. To date, the von Thunen land use model is considered to be the reference land use model and is widely applied in land use modelling techniques (von Thunen, 1966). However, of late land use/land cover change models have followed a multidisciplinary approach involving various fields such as geography, environment and ecosystem sciences, civil engineering, biophysics, chemistry, remote sensing and mathematics. Emergence of an advanced LULCC model started with a collaborative project initiated in the year 1995 with support from the International Geosphere-Biosphere Programme (IGBP) and the International Human Dimensions Programme on Global Environmental Change (IHDP) (Basse et al., 2014). Models have received attention from a number of academicians and scientists mainly to fill the research gap in the existing techniques such as

1. Models to help in exploring the dynamic changes with various factors in consideration
2. Spatio-temporal assessment of future trends or potential of land use changes
3. Facilitating exchange of information not only between researchers and academicians but also between policymakers and planners to arrive at significant actions required to achieve sustainable development goals

Currently, there is a wide spectrum of land use change models available, for instance, the cellular automata–Markov chain (CAMC), SLEUTH, CLUE-S and CA-ANN, that have been designed at both the local and global scale. Apart from these, simulation using agents is also gaining popularity. The aim of an agent-based model (ABM) is to create an environment where interactions are programmed to be self-constrained or, in other words, actions that can be controlled on their own. An agent is therefore autonomous, functioning independently, interacting socially with other agents in a specified environment, while following explicit goals that are responsible for the agent's behaviour (Macal & North, 2009; Franklin & Graesser, 1996). Axtell (2000) discusses the need of agents in situations where it is important to adapt and change behaviours, and to learn and engage in dynamic relationships with other agents, adding a spatial component to identify behaviour and interactions of agents in a real-world environment. An autonomous active agent is defined by the following characteristics:

• An environment, which is a dynamical system where agents operate and interact with other agents, and changes over time.
• The sensing capabilities of an agent are defined with respect to its sensors. Sensing of an agent may be active or passive, or internal or external depending on the environment.

- The action of an agent causes significant change in the current state and is responsible for behaviour in the future. Agents are often goal-oriented.
- Interaction, meaning agents have extensive ability to communicate with other agents.
- Drives are also considered as preferences. They can be observed because of the relationship between functional modules of an agent. The agent may have a single drive or multiple drives.
- An agent's action selection guides future predictions and initiates actions accordingly.
- With mobility, agents have all the freedom to roam around within a specified environment. Taking advantage of this, agents are often interspersed and juxtaposed to interact permitting a vast range of applications.
- With self-learning, agents possess a unique characteristic of adaptive or complex adaptive systems where agents alter their state depending on a previous state (Holland, 1995).

Figure 1.5 illustrates the basic structure of an agent. An agent interacts with landscape leading to four possible relations: self-influencing behaviour of agents, self-influencing state of a landscape at any given time, agents affecting landscape and vice versa. Apart from these relations, an agent might also interact with other agents and environments, while landscape is dependent on external variables or could be influenced by other factors (Heppenstall et al., 2012).

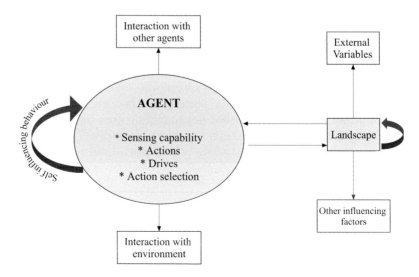

FIGURE 1.5
Structure of an agent. (Adapted from Batty, 2005.)

TABLE 1.2

Advantages and Comparison of ABM versus Traditional Modelling
Method

Traditional Modelling	ABM
Deterministic (limited to one feature)	Stochastic (multiple features)
Allocative (top-down approach)	Aggregative (bottom-up approach)
Equation based	Adaptive agents
No explanations	Explanatory in nature
Considers few parameters	Considers many parameters
Spatially coarse	Spatially explicit
Works in given environment	Environment is created

The major advantages of an ABM over traditional modelling techniques
are threefold:

1. Captures emerging spatio-temporal footprint of land use change
2. Provides the domain-specific environment of the region under
   investigation
3. Highly flexible, for developing geospatial models in particular

Apart from these advantages, Bernard (1999) lists a few characteristics of
the traditional modelling method versus ABM (listed in Table 1.2). There are
challenges that need to be addressed when it comes to land use change models such as there is no particular framework available for systematic assessment of land use/land cover change prediction and hence its validation
considering the decision-making process and authenticity of interactions
between various actors in a system.

Another major challenge lies in resolution of data, availability and trade-
offs between different resolutions. Since modelling results directly depend
on various aspects of remote sensing resolution, predicting quantities of
change would not be reliable if spatial patterns are not considered under the
lens of resolution. Land use change is dependent on land management policy framed by local, regional, national and global agencies. Therefore, there
is a need to integrate these factors along with human–environment interaction to achieve desirable results.

1.7 Indian Context

India has a population of 1.21 billion (2011) of which 32% live in urban areas.
The population density has increased from 117 per sq km in 1951 to 368 per

sq km in 2011. It is estimated that urban areas contribute more than 60% of national output (GOI, 2010) and about 80% of total tax revenue (Shanke et al., 2010). It is clearly evident that increasing urbanisation is the prime cause of higher national GDP in the Indian economy. The country in this regard has been witnessing a rapid LULCC leading to concentrated growth of paved structures with a steep increase in land prices and later to a fragmented outgrowth from the city boundaries towards the urban–rural side. This fragmented outgrowth often results in a phenomenon called sprawl. Sprawl may be defined as regions and settlements that are located near a city and/or the outskirts of the growing urban area with very low to low-density developments that would transform into high-density developments in a few years depending on the economy and other major factors of the city. These are connected to the urban core without proper recognition from city development authorities or planners. These regions are devoid of essential infrastructure and basic amenities like water, transport connectivity, sanitation and electricity. Such expansions are normally and extensively associated with growth that is uncoordinated and could form a crisis for sustainable city development and future planning of city growth.

Urbanisation is one of the facets of rapid LULCC, also taking place under significantly different institutional conditions in different Indian states, and has been subject to various policies and planning strategies. Diverse trajectories of land reform across the states have continued to impose a variety of conditions for rural land acquisition and development. Urbanisation has also taken place under a range of local planning regimes, from integrated, planned townships to unplanned or organic growth. Rapid urban growth in the Indian scenario can be attributed to rural to urban migration, reclassification of cities, improved health facilities, etc. in urban areas. With the commencement of various urban development policies and schemes such as smart cities, AMRUT (Atal Mission for Rejuvenation and Urban Transformation), JNNURM (Jawaharlal Nehru National Urban Renewal Mission), NERUDP (North Eastern Region Urban Development Programme) and Capacity Building for Urban Local Bodies have promoted migration to core, transition or hinterland areas. Comparing census statistics, the highest urban population percentages were recorded in Maharashtra (35%) and Tamil Nadu (48.5%) in the years 1981 and 2011, respectively. Figure 1.6 shows population trends of major Indian cities for the years 1991, 2001 and 2011. India already has three megacities (Mumbai, Delhi and Kolkata urban agglomeration) with populations greater than 10 million and 42 conurbation areas or million-plus cities (Census of India, 2011). By 2025, Asia alone will have at least 28 megacities. The rate of urbanisation in India shows moderate progression (24% in 1981, 26% in 1991, 28% in 2001 and crossing 31.16% as per 2011 census).

Large-scale LULCC is linked to these shifts, such as the loss of forests to meet the urban demands of fuel, timber and land, which have led to changes in the ecosystem structure, affecting its functioning and thereby threatening sustainable development. Cities are often regarded as the engines of

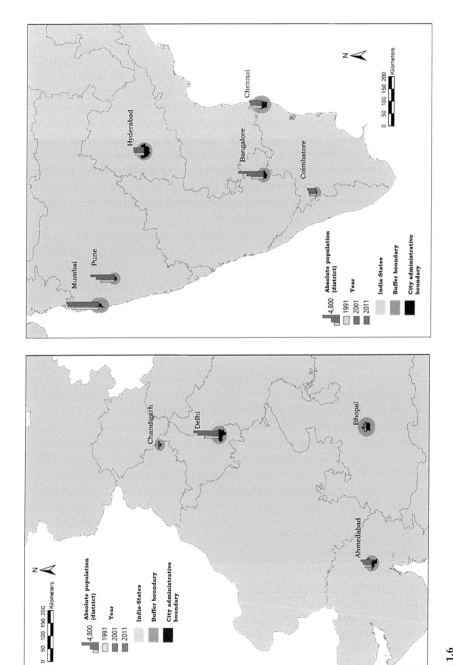

FIGURE 1.6
North (left) and South (right) India cities and populations in 1991, 2001 and 2011. (Data from Census of India.)

economic growth. This necessitates that cities get smarter to handle this large-scale urbanisation and find new ways to manage complexity, increase efficiency and improve quality of life. The design of smart cities requires an understanding of spatial patterns of urbanisation to implement appropriate mitigation measures. This requires spatial information to visualise the patterns of urbanisation and predict the likely changes with the implementation of decisions (what-if scenarios) by city administrators. Smart cities would be self-reliant and self-sufficient systems with provision for good infrastructure such as water, sanitation and reliable utility services; health care; transparent government transactions; and various citizen-centric services.

1.8 Concluding Remarks

Understanding spatial patterns of landscape and landscape structure is necessary for assessing impacts on environment and planning development through the lens of sustainability. This chapter highlights the importance of understanding landscape structure, usage of geospatial technology in addressing landscape change issues, spatio-temporal aspects of landscape changes, strategies used in land use assessment such as classification systems, models developed to understand land use change and their advantages and limitations over traditional modelling techniques, and global as well as Indian landscape changes. Also, the chapter describes the fundamentals of remote sensing, GIS and usage of satellite-based remote sensing data to comprehend realistic growth through data visualisation, which forms the fundamental knowledge base for other chapters of this book such as land surface temperature estimation thus relating issues of greenhouse gas (GHG) emissions; measuring pattern change using spatial metrics, planning and prediction of land use changes; and status of sustainable development goals in India, with a special focus on disaster management. There lies an opportunity for research in landscape planning and management since governments in developing countries, for instance, India, are unable to match people-to-land ratio with concerns towards the environment and sustainability. Research in this field requires a multidisciplinary perspective with collaboration from engineers, ecologists, geospatial analysts and policymakers to provide region-specific landscape solutions involving environmental, socio-economic, cultural and political aspects.

References

Anderson, J. R. 1976. *A land use and land cover classification system for use with remote sensor data*. US Government Printing Office.

Antrop, M. 1998. Landscape change: Plan or chaos? *Landscape and Urban Planning*, *41*(3–4), 155–161.

Axtell, R. 2000. Why agents? On the varied motivations for agent computing in the social sciences. *Center on Social and Economics Dynamics - The Brookings Institution*, *17*, 1–23.

Axtell, R., & Epstein, J. 1994. Agent-based modeling: Understanding our creations. *The Bulletin of the Santa Fe Institute*, *9*(4), 28–32.

Basse, R. M., Omrani, H., Charif, O., Gerber, P., & Bódis, K. 2014. Land use changes modelling using advanced methods: Cellular automata and artificial neural networks. The spatial and explicit representation of land cover dynamics at the cross-border region scale. *Applied Geography*, *53*, 160–171.

Batty, M. 2005. Agents, cells, and cities: New representational models for simulating multiscale urban dynamics. *Environment and Planning A*, *37*(8), 1373–1394.

Bernard, R. N. 1999. Using adaptive agent-based simulation models to assist planners in policy development: The case of rent control. *Santa Fe Institute Working Papers*, *7*(52), 18.

Census of India. 2011. *Report on post enumeration survey*. Accessed on November 29, 2018. Retrieved from http://censusindia.gov.in/

Chang, K. T. 2015. *Introduction to geographic information systems*, 8th edition. McGraw-Hill Higher Education publication.

Congalton, R. G. 1991. A review of assessing the accuracy of classifications of remotely sensed data. *Remote Sensing of Environment*, *37*(1), 35–46.

Cotton, W. R., & Pielke Sr, R. A. 2007. *Human impacts on weather and climate*. Cambridge University Press.

Foley, J. A., DeFries, R., Asner, G. P., Barford, C., Bonan, G., et al. 2005. Global consequences of land use. *Science*, *309*(5734), 570–574.

Forman, R. T. T. 1995. *Land Mosaics: The ecology of landscapes and regions*. Cambridge: Cambridge University Press.

Franklin, S., & Graesser, A. 1996. Is it an agent, or just a program? A taxonomy for autonomous agents. In *Proceedings of the Third International Workshop on Agent Theories, Architectures, and Languages* (pp. 21–36).

Gardner, R. H., Lookingbill, T. R., Townsend, P. A., & Ferrari, J. 2008. A new approach for rescaling land cover data. *Landscape Ecology*, *23*(5), 513–526.

Government of India (GOI). 2010. *Mid-term appraisal of the eleventh five-year plan 2007-2012, Planning Commission, GOI*. Accessed on December 4, 2018. Retrieved from http://planningcommission.nic.in/

Heppenstall, A. J. J., Crooks, A. T., See, L. M., & Batty, M. 2012. *Agent-based models of geographical systems*. Netherlands: Springer.

Holland, J. H. 1995. *Hidden order: How adapation builds complexity* (Vol. 75). Basic Books.

Houet, T., Verburg, P. H., & Loveland, T. R. 2009. Monitoring and modelling landscape dynamics. *Landscape Ecology*, *25*(2), 163–167.

Jensen, J. R., & Cowen, D. C. 1999. Remote sensing of urban/suburban infrastructure and socio-economic attributes. *Photogrammetric Engineering and Remote Sensing*, *65*(5), 611–622.

Kabat, P., Claussen, M., Dirmeyer, P. A., Gash, J. H., de Guenni, L. B., et al. (Eds.). 2004. *Vegetation, water, humans and the climate: A new perspective on an internactive system*. Springer Science & Business Media.

Kasanko, M., Barredo, J. I., Lavalle, C., McCormick, N., Demicheli, L., et al. 2006. Are European cities becoming dispersed? A comparative analysis of 15 European urban areas. *Landscape and Urban Planning*, 77(1–2), 111–130.

Lausch, A., Blaschke, T., Haase, D., Herzog, F., Syrbe, R.-U., et al. 2015. Understanding and quantifying landscape structure – A review on relevant process characteristics, data models and landscape metrics. *Ecological Modelling*, 295, 31–41.

Lillesand, T. M., Kiefer, R. W., & Chipman, J. W. 2012. *Remote sensing and image interpretation*, 6th edition. John Wiley and Sons.

Macal, C. M., & North, M. J. 2009. Agent-based modeling and simulation. In *Proceedings of the 2009 Winter Simulation Conference* (pp. 86–98).

National Research Council (NRC). 2005. *Radiative forcing of climate change: Expanding the concept and addressing uncertainties* (p. 208). Committee on Radiative Forcing Effects on Climate Change, Climate Research Committee, Board on Atmospheric Sciences and Climate, Division on Earth and Life Studies. Washington, DC: The National Academies Press.

Parker, D. C., Manson, S. M., Janssen, M. A., Hoffmann, M. J., & Deadman, P. 2003. Multi-agent systems for the simulation of land-use and land-cover change: A review. *Annals of the Association of American Geographers*, 93(2), 314–337.

Paudel, S., & Yuan, F. 2012. Assessing landscape changes and dynamics using patch analysis and GIS modeling. *International Journal of Applied Earth Observation and Geoinformation*, 16, 66–76.

Qiuying, L., Chuanglin, F., Guangdong, L., & Zhoupeng, R. 2015. Quantitative measurement of urban expansion and its driving factors in Qingdao: An empirical analysis based on county unit data. *Journal of Resources and Ecology*, 6(3), 172–179.

Ramachandra, T. V., Bharath, H. A., & Sanna, D. D. 2012. Insights to urban dynamics through landscape spatial pattern analysis. *International Journal of Applied Earth Observation and Geoinformation*, 18, 329–343.

Ramachandra, T. V., Bharath, H. A., & Vinay, S. 2013. Land use land cover dynamics in a rapidly urbanizing landscape. *SCIT Journal*, 13(1), 1–13.

Sankhe, S., Vittal, I., Dobbs, R., Mohan, A., Gulati, A., et al. 2010. *India's urban awakening: Building inclusive cities sustaining economic growth*. McKinsey Global Institute. Accessed on December 16, 2018. Retrieved from https://www.mckinsey.com/

Saunders, D. A., Hobbs, R. J., & Margules, C. R. 1991. Biological consequences of ecosystem fragmentation: A review. *Conservation Biology*, 5(1), 18–32.

Sudhira, H. S., Ramachandra, T. V., & Jagadish, K. S. 2003. Urban sprawl pattern analysis using GIS. *CES technical report no. 99*. Accessed on December 07, 2018. Retrieved from http://wgbis.ces.iisc.ernet.in/

Turner, M. G. 1987. Spatial simulation of landscape changes in Georgia: A comparison of 3 transition models. *Landscape Ecology*, 1(1), 29–36.

von Thünen, J. H. 1966. Der isolierte staat in beziehung auf landwirtschaft und nationalokonomie. In Hall, P. (Ed.), *von Thünen's isolated state*. Oxford: Pergamon.

Wu, W., Zhao, S., Zhu, C., & Jiang, J. 2015. A comparative study of urban expansion in Beijing, Tianjin and Shijiazhuang over the past three decades. *Landscape and Urban Planning*, 134, 93–106.

2

Quantification of Greenhouse Gas Footprint

CONTENTS

2.1 Introduction 19
2.2 GHG Footprint 24
2.3 GHG Footprint: Indian Scenario 26
2.4 Method for Estimating GHG Emissions... 29
2.5 Case Study of Delhi 33
2.6 Conclusion 35
References... 35

2.1 Introduction

Greenhouse gases (GHGs) include species of carbon and nitrogen that are occurring naturally or induced by human activities that absorb and emit energy within the thermal infrared region of the electromagnetic (EM) spectrum (IPCC, 2014) (Figure 2.1). These gases allow shortwave radiations including the visible and ultraviolet (UV) portion of the EM spectrum to reach Earth's surface unobstructed, while the ozone layer filters harmful UV radiations. The filtered shortwave energy reaches Earth's surface, where part of it is absorbed and heats Earth's surface and the remaining reflects back in the form of long-wave infrared energy. GHGs have a tendency to trap these long-wave radiations by the process of absorption, forming an invisible blanket around the earth that doesn't allow sudden changes in the earth's temperatures (NOAA, 2019a). This phenomenon of keeping the lower atmosphere of the earth warm is termed the greenhouse effect (Figure 2.2).

GHGs play a vital role for survival on Earth, but over the past century, there has been a significant rise in atmospheric concentration of these gases due to escalating human activities in combination with the increased demographic pressure causing an alteration in global temperature leading to changes in the micro- and macro-climate. Emissions from activities due to industrialisation and urbanisation are some factors that cause magnification in atmospheric concentrations of GHGs (Ramanathan & Parikh, 1999). The atmosphere consists of nitrogen (78.1%), oxygen (21%) and trace gases that are present in small quantities. GHGs including carbon dioxide (CO_2), methane (CH_4), nitrous oxide (N_2O), sulphur hexafluoride (SF_6), water vapour

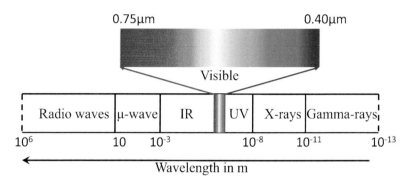

FIGURE 2.1
Electromagnetic spectrum.

(H_2O), ozone (O_3), chlorofluorocarbons (CFCs) and hydrofluorocarbons (HFCs) are part of these trace gases that have strong infrared absorption bands and significantly affect the atmospheric thermal structure (Wang et al., 1976). GHG emissions are measured in terms of CO_2e (carbon dioxide equivalent) out of which CO_2 is the most dominant and contributes around 77% (IPCC, 2014), followed by CH_4 and N_2O.

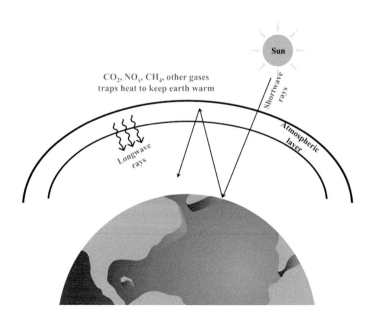

FIGURE 2.2
Illustration of greenhouse effect.

1. *Carbon dioxide.* There has been persistent rise in CO_2 levels. Before 1958, CO_2 levels were identified by the air enclosed in polar ice caps (Berner et al., 1980). It has been established that historically there has been rise and fall in concentration of CO_2, but it had never crossed the 300 ppm (parts per million) mark until the 1950s (www.climate. gov). During the ice age, the CO_2 level was 200 ppm and during the warmer interglacial period it was 280 ppm (Indermuhle et al., 1999). It was only in 2013 when the level crossed 400 ppm and it was 405 ppm in 2017, which is the highest in the past 800,000 years. More than 3 million years ago, the atmospheric concentration of CO_2 was high and the temperature along with sea level was also higher than the pre-industrial era (NOAA, 2019b). There have been both natural and anthropogenic sources of CO_2 emissions, wherein natural sources comprise ocean release, respiration, decomposition, etc., whereas emissions due to industrial processes and production, deforesta- tion, land use changes, burning of fossil fuels, natural oil and coal, etc., are the human-induced sources (Neftel et al., 1982; WYI, 2019). Plants, algae (in waterbodies and oceans, lentic and lotic) and soil sequester carbon naturally, but now these carbon sinks are unable to maintain equilibrium with the escalated GHG footprint due to burgeoning anthropogenic activities as a result of globalisation.

2. *Methane.* Methane is the second most important GHG after CO_2, as its global warming potential (GWP) for a time horizon of 20 years is 56 times, while for 100 years is 21 times that of CO_2 (UNCC, 2019). The GWP for CH_4 includes unintended alterations in tropospheric ozone production and stratospheric water vapour production. As of 2005, its concentration increased by around 2.5 times (1774 parts per bil- lion [ppb]) from pre-industrial levels (715 ppb) (Boucher et al., 2009). As of November 2018, the concentration had increased to 1867.2 ppb (NOAA, 2019c). Major anthropogenic sources of GHG include fos- sil fuel production and usage, fermentation in cattle, cultivation of rice (transplantation), manure production, waste management and biomass burning. Anthropogenic sources account for around 50% of the total methane emission. Other natural sources of CH_4 include oceans, wildfires, permafrost, wetlands and soil. The biggest sink of CH_4 is atmosphere, as it reacts with OH^- ions and forms CO_2 and H_2O. The global monthly mean of methane can be referred from www.esrl.noaa.gov/.

3. *Nitrous oxide.* Nitrous oxide is another prominent GHG that has a GWP of 280 times over a time horizon of 20 years and 310 times over a time horizon of 100 years when compared with CO_2 (UNCC, 2019). In the previous 800,000 years, the levels crossed 300 ppb only twice (303 ppb 786,500 years ago and 302 ppb 330,400 years ago), but since the 18th century there has been a steep rise in the concentration.

The current level of N_2O as of January 2019 was about 332.75 ppb (Two Degree Institute, 2019). N_2O emissions have both natural as well as anthropogenic sources. Natural sources contribute around 62% of the total emissions and are formed due to numerous microbial activities in soil (nitrification) and ocean spray (WYI, 2019). Anthropogenic sources include agricultural activities, fossil fuels combustion, human and animal sewage, and biomass burning. The atmosphere and soil (denitrification) act as major sinks for nitrous oxide (GHG Online, 2019). The global level of N_2O can be referred from www.n2olevels.org/.

4. *Fluorinated gases.* These include gases such as sulphur hexafluoride, CFCs, HFCs and perfluorocarbons. These gases contribute a small fraction of the total GHGs but due to their higher GWP, they play an important role in the overall estimation. Their GWP ranges from 460 to 16,300 times that of CO_2 for a time horizon of 20 years and 150 to 23,900 times that of CO_2 for a time horizon of 100 years (UNCC, 2019). All these gases are synthetic and created during manufacturing and production of metals and semiconductors, and other major industrial processes. Refrigeration and air conditioning contribute to more than three-fourths of the total emission followed by foams and aerosols (WYI, 2019).

Estimating and monitoring the sources and sinks of these GHGs is very vital to maintaining the equilibrium, as it affects the atmosphere, land surface, waterbodies, ice caps and snow, etc. (Le et al., 2007; Maslin, 2008). The increasing concentration of GHGs is a major reason for global warming that melts freshwater ice, increasing the sea level. Ocean acidification is another impact of the increased level of CO_2 that affects marine biodiversity. These are due to the anthropogenic sources of GHGs causing Earth's climate change (McMichael et al., 2006). The recent rise in temperatures leading to thermal instability, increased weather extremes, rise in cases of droughts and floods, increased cases of infectious diseases, and thermal discomfort are inferred from human-induced emissions. The global mean ambient air temperature has increased by 0.5°C when compared with the 19th century. According to the Fourth Assessment Report of the Intergovernmental Panel on Climate Change (IPCC), the atmospheric concentration of GHGs has increased and triggered a rise in global mean temperature. This initiated an increase in the number of hot days, hot nights and decreased rainy days. Continental precipitation has increased in eastern parts of America (North and South), northern Europe and Asia, and decreased in Africa, the Mediterranean and southern Asia (Pachauri & Chand, 2008). As per the latest prediction (Wigley & Raper, 2001), global temperature might increase 1–6°C by 2100 with alterations in precipitation, evapotranspiration and runoff, impacting water supply and biodiversity. Abrupt changes in the climate threaten water and food security throughout the globe. Temperature alterations propagate gradually into rocks beneath the surface of the earth changing the ambient thermal regime and damaging the natural subsurface balance (Pollack et al., 1998). The first decade of 21st century had the warmest global temperatures recorded, starting from the late 19th century (NOAA, 2018).

Considering the threats of global warming and consequent changes in the climate, there are attempts to evolve strategies to mitigate the GHG footprint. Deliberations on climate change have been done through international conferences organised by the United Nations Framework Convention on Climate Change (UNFCCC). Since 1992, environmental-related treaties have been signed to combat climate change for a sustainable future. Numerous scientific societies including the American Association for the Advancement of Science (AAAS), American Meteorological Society (AMS), American Geophysical Union (AGU), American Medical Union (AMU), science academies (US National Academy of Sciences), and US government agencies (US Global Change Research Program) have stated that there is sufficient and strong scientific evidence proving that climate is changing globally due to increased anthropogenic interventions causing a serious threat to society (NASA, 2018; Omenn et al., 2006; AMS, 2012).

Another major step to mitigate these issues was through establishment of the IPCC in 1998 by the United Nations. In 1992, at the first international conference, the Rio Convention was signed by 154 nations to reduce the emissions within 10 years to avoid global warming. In 1997, through the Kyoto Protocol there was an agreement among the nations to reduce emissions by 5.4% by 2010 (Marshall, 2009). In 2015, at the climate change convention, most of the participating nations resolved to restrict the temperature rise to 1.5–2°C. This agreement was ratified by 195 nations (UNCC, 2018). These initiatives necessitated GHG accounting at micro and macro levels. Some of the predicted negative impacts of climate change on the environment and hominids by 2100 (UN Development Program, 2018) are

- The temperature will soar more than 1.5°C.
- Natural systems on Earth, permafrost and glaciers will be affected.
- Floods and other natural disasters will affect more than 2 billion people.
- One-third of world population will be affected due to sea level rise.
- Adaptation measures for a rise of 2°C will cost around 70 billion–100 billion USD per year.
- Every continent will be affected causing danger to all living beings.

2.2 GHG Footprint

The GHG footprint is an aggregate of all emissions from greenhouse gases considering all sectors of anthropogenic activities. It is expressed in terms of CO_2 equivalent (CO_2e), and as per the IPCC it is a unit for comparing the radiative force using 100 year GWP of a GHG to that of CO_2. It is quantified by the multiplying the mass of a gas with its respective GWP (BSI, 2011). The GWP of important GHGs are as shown in Table 2.1.

Human communities across the globe, perceiving the imminent threats of changes in climate, have started exploring strategies to mitigate GHG emissions responsible for global warming, and this has been one of the biggest

TABLE 2.1

Global Warming Potential of Greenhouse Gases

Species	Chemical Formula	Global Warming Potential (Time Horizon)		
		20 Years	100 Years	500 Years
Carbon dioxide	CO_2	1	1	1
Methane	CH_4	56	21	6.5
Nitrous oxide	N_2O	280	310	170
HFC-23	CHF_3	9100	11700	9800
HFC-32	CH_2F_2	2100	650	200
HFC-41	CH_3F	490	150	45
HFC-43-10mee	$C_5H_2F_{10}$	3000	1300	400
HFC-125	C_2HF_5	4600	2800	920
HFC-134	$C_2H_2F_4$	2900	1000	310
HFC-134a	CH_2FCF_3	3400	1300	420
HFC-152a	$C_2H_4F_2$	460	140	42
HFC-143	$C_2H_3F_3$	1000	300	94
HFC-143a	$C_2H_3F_3$	5000	3800	1400
HFC-227ea	C_3HF_7	4300	2900	950
HFC-236fa	$C_3H_2F_6$	5100	6300	4700
HFC-245ca	$C_3H_3F_5$	1800	560	170
Sulphur hexafluoride	SF_6	16300	23900	34900
Perfluoromethane	CF_4	4400	6500	10000
Perfluoroethane	C_2F_6	6200	9200	14000
Perfluoropropane	C_3F_8	4800	7000	10100
Perfluorobutane	C_4F_{10}	4800	7000	10100
Perfluorocyclobutane	$c\text{-}C_4F_8$	6000	8700	12700
Perfluoropentane	C_5F_{12}	5100	7500	11000
Perfluorohexane	C_6F_{14}	5000	7400	10700

Source: United Nations Climate Change, https://unfccc.int/.

scientific and developmental challenges (Fan et al., 2007). There are multiple factors such as technology, infrastructure, regulatory framework, policies and lifestyle that need to be immensely modified for transition towards a low carbon economy (Sovacool & Brown, 2010). Moreover, from the past few years, there has been a growing awareness about climate change by the general public and government officials, and the GHG footprint and climate change are perceived as matters of great concern. Many cities across the globe are now setting up organisations and stations to check their GHG footprint, ambient air quality and temperature, so as to maintain a healthy environment and combat against global climate change. Also industries are being transferred to other cleaner industries so as to let the economic development continue at an advanced level along with greener technology. There is a necessity of targets being set with time frames that can aid decision makers at local, national and international levels to create sustainable frameworks and policies. Carbon intensity, one of the vital indicators that can help in measuring carbon dioxide emissions per unit of economic activity, is usually measured as GDP (gross domestic product). Carbon intensity and energy efficiency are closely correlated to the economy of a country. If a country's economy improves or its overall emissions are reduced, then the carbon intensity value for that country decreases. Figure 2.3 demonstrates the carbon intensity of major countries across the world. India has managed to reduce its carbon intensity marginally in recent years despite coal being the major fuel in electricity production. This could be explained by improvement in technology and strong wind penetrations.

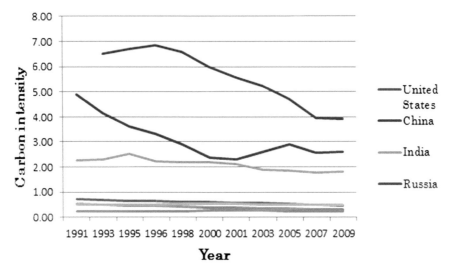

FIGURE 2.3
Carbon intensity of major carbon players across the globe. (Ramachandra et al., 2015.)

It is necessary to quantify the GHG footprint, as it has potential to reduce the impact of climate change by structural changes in various sectors (with improvements in energy efficiency, decarburisation through adoption of carbon neutral renewable sources of energy, etc.) and more important through enhanced awareness amongst local groups, individuals and government officials. It has the capability to drive discussions and provide valuable inputs to decision makers, stakeholders, local municipalities and general public that can lead to significant transformation in the framework. Sector-wise estimation can help in understanding the area with the highest emissions and priorities can be set according to that.

2.3 GHG Footprint: Indian Scenario

There have been numerous studies related to climate change carried out in India, but an issue is the lack of proper documentation. The first report that had complete inventory of GHG emissions for 1990 and projection for 2020 was published under the Asia Least Cost Greenhouse Gas Abatement Strategy Project. Another report was submitted to UNFCCC in 2004 for fulfilment of the responsibility that covered assessment of GHG emissions from various sectors as per IPCC guidelines. It is now a necessity to estimate sector-wise emissions at regular time intervals for formulating appropriate policies. Some of the major sectors contributing to GHG emissions are as shown in Figure 2.4.

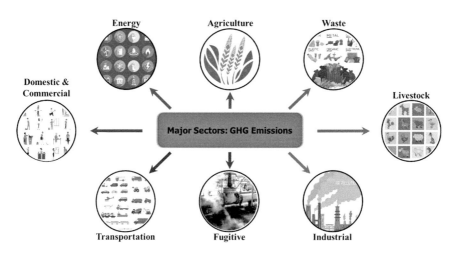

FIGURE 2.4
Major sectors of GHG emissions.

1. *Energy.* The energy sector plays a crucial role in the economy while catering to a wide range of applications from rural households to large industries. Electricity growth in India has overtaken the economic growth in recent times (SSEF, 2017). In terms of production, India stands in the third position globally, while in case of consumption, it is in fourth position. The installed capacity by the Indian power sector has grown to 326,848.54 MW (as of March 2017) from just 1713 MW in 1950. Coal contributes to almost 60% of the total production followed by hydro, gas, nuclear and diesel (Central Electricity Authority, 2018). Combustion of fuel for electricity generation emits CO_2, NO_x, SO_x and inorganic particulates (suspended particulates, PM_{10}, $PM_{2.5}$, PM_1 and fly ash). The energy sector has been a dominant source of emissions in many Indian cities including Delhi.

2. *Domestic and commercial.* The domestic sector is a major sector that contributes a considerable amount of emissions from lighting, use of various household appliances and fuel burned for cooking. The domestic sector consumes the largest amount of fuel after industry. The commercial sector mostly uses oil and natural gas, and the domestic sector uses liquefied petroleum gas (LPG), wood and coal as fuel. Gross CO_2 emissions in this category are due to inefficient or unburnt fuel being disseminated with the gases. In rural India, biomass burning or inefficient cooking devices, and in urban areas, inefficient design of buildings with glass facades and Western design are creating perilous situations in the tropical regions near the equator as they require more energy-intensive devices such as air conditioners to moderate temperature (Jayalakshmi, 2015). The overall CO_2 emissions from the domestic and commercial sectors in 2007 was around 140 million tons of CO_2e (INCCA, 2010).

3. *Transportation.* The transportation sector is responsible for around 21% of global GHG emissions and is the third-ranked contributing sector in India (TWB, 2014; Singh et al., 2019). The transportation sector has shown exponential growth (Statista, 2019), for example, the total number of vehicles has increased from 81.5 million to 230 million during 2005 to 2016 in India. This is contributing to GHGs such as NO_x, SO_x, CO, CO_2, ammonia, sulphur and particulate matters. Emissions from the transportation sector have doubled since 1970s. Major sources of energy in this sector are gasoline (petroleum and diesel), coal, compressed natural gas (CNG) and electricity. The major constituents of mode of transport are road (80%), railways (13%) and aviation (6%). Globally, GHG emissions from this sector have increased to 7 Gt CO_2e from 2.8 Gt CO_2e over a time period of four decades (Sims et al., 2014).

4. *Industrial.* This sector contributes to approximately one-third of the total global anthropogenic GHG emissions and 25% of total

emissions in India. The primary industries involved are manufacturing, chemicals, petrochemicals, cement, iron and steel and paper and pulp. As a result of industrialisation, the period from 1850 to 1960, there was a constant growth of emissions around the world (mostly in the United States). Since that time Asian countries have shown an enormous increasing trend. The major emission from this sector is CO_2. In India, the cement industry contributes significantly, followed by iron and steel, aluminium, paper and glass (GGW, 2019). The industrial capacity of India is expected to increase and contribute 25% of GDP by 2030 (Pappas & Chalvatzis, 2017).

5. *Agriculture.* More than half of the Indian population depends on the agriculture sector for its livelihood, and this sector contributes to 17% of total GDP (Arjun, 2013). India has the highest net cropping area across the globe and exports more than USD 40 billion worth of agricultural products. Agrarian activities contribute to direct release of GHGs from numerous sources. Major sources are CH_4 via rice cultivation, N_2O from fertilisers rich in nitrogen and CO_2 release from tube wells and pumps. Another major release of GHGs from the agricultural sector is open stubble burning. Agricultural by-products are fed to cattle and also used as domestic fuel; the remainder is burnt in the field that causes increased particulate matters in addition to GHG emissions. Also, cultivation of rice paddies leads to release of additional sources of methane as a result of anaerobic decomposition (Ramachandra et al., 2015).

6. *Waste.* Waste management is one of the key sectors, contributing about 4% of India's total GHG emissions. CH_4 is the major GHG from this sector as it is a by-product of anaerobic decomposition of organic fraction of solid waste. Another source of CH_4 is indiscriminate disposal of domestic and industrial wastewater. Methane production can be determined by estimating the amount of biodegradable/organic matter present in wastewater and can be expressed as biological oxygen demand (BOD) and chemical oxygen demand (COD). N_2O is another emission from organic waste due to the presence of protein. In the case of waste dumps, exposed waste emits CO_2 in addition to CH_4. As a result of increased waste generation, overall emissions from the waste sector increased by 36% in 2013 compared to 2005 levels. A major contributor of the waste sector is domestic wastewater (59.8%), followed by industrial wastewater (23.5%) and solid waste disposal (16.7%) (Garg et al., 2011; ICLEI, 2017).

7. *Livestock.* Food production in terms of intensive livestock nurturing for meat and dairy is a vital source of GHG production. About half of livestock emissions are in the form of methane and the rest is shared by N_2O and CO_2. Livestock from India contributed to around 15.3 million tons of methane emission in 2012 (SciDev, 2018). Globally,

emissions from livestock contribute 14.5% of total anthropogenic GHG emissions. Methane is the major emission owing to this sector and the major sources includes enteric fermentation as a result of the digestive process of ruminants (44%), fodder production, and processing and animal waste management (41%), followed by manure storage and processing (10%), and the remaining can be ascribed to energy consumption in transportation of products. Beef manufacturing contributes the most emissions of about 41%, followed by cattle milk (20%), pig meat (9%), buffalo milk and meat (8%), chicken meat and eggs (8%) and small ruminants (6%), and the rest can be attributed to other poultry species and non-edible products (FAO, 2019). As per the estimate, 10–12% of total emission is due to beef in the US, and replacement with chicken or any other cattle could reduce 80% of emissions from this sector (Carbon Brief, 2016).

8. *Fugitive.* This is one of the neglected sectors for GHG emissions. Fugitive emission means release of GHGs indirectly into the atmosphere as a result of manufacturing or working processes, usage of equipment and any undesirable leakages. This sector includes sources such as refrigeration, air conditioning, heat pumps, cold-storage warehouses, fixed and portable fire suppression equipment, industrial leakages, manufacturing, testing and laboratory discharges. This sector contributes 40 Gt CO_2e and is predicted to rise in the future (Ritchie & Roser, 2019).

2.4 Method for Estimating GHG Emissions

Often, GHG estimation is performed by combining all the sectors, but it might provide inaccurate results as emissions from different sectors vary with diverse parameters. This type of estimation tends to result in erroneous values, thus not being able to serve the purpose of formulating strategic plans and policies for mitigation measures. The sector-wise GHG footprint for seven Indian megacities was estimated considering all the sectors mentioned in the previous section (Ramachandra et al., 2015)

GHGs were estimated and converted into CO_2e by using GWP, as per Table 2.1. Total emissions of GHG from all its sources can be quantified as

$$\text{Emission}_{\text{Gas}} = \sum_{\text{Category}} A \times \text{EF} \qquad (2.1)$$

where $\text{Emission}_{\text{Gas}}$ is the emission of gas from its sources, A is the amount of individual source category utilised that generates emissions of the gas under consideration and *EF* is the emission factor of gas under consideration.

The emission factor can be defined as the characteristic value that relates the quantity of GHG with an associated activity responsible for release of that specific GHG. The IPCC defines the emission factor as the average rate of emission of a particular GHG for a given source associated to its activity. It is usually expressed as a ratio of weight of GHG to unit weight/volume/distance/duration. This quantification varies as per different sectors, thus sector-wise estimation should be performed as follows:

1. *Energy.* Emissions in this sector are due to the combustion of fossil fuels in thermal power plants. Major emissions include CO_2, SO_x, NO_x, airborne inorganic particulates, particulate matter and fly ash. Total emissions from a single source in this sector is estimated as per the following equation:

Emission = Fuel consumption × Net calorific value of that fuel × Emission factor   (2.2)

Similarly, emissions for all the fuels such as coal, natural gas, diesel oil, naphtha, low sulphur heavy stock, low sulphur fuel oil, residual fuel oil and heavy fuel oil should be estimated with their respective calorific value and emission factor.

As per the IPCC, an alternate basic equation (CEEW, 2019) can be used to estimate emissions from a gas:

$$E_{Gas} = A_{fuel} \times CV_{Unit} \times CV_{fuel} \times EF_{Gas} \times GWP_{Gas}$$ (2.3)

where E_{Gas} is the emission of gas (tonnes), A_{fuel} is the activity data of fuel (litres/kg/tonnes), CV_{Unit} is the conversion factor (activity data to tonnes), CV_{fuel} is the calorific value of fuel, EF_{Gas} is the emission factor of GHGs, and GWP_{Gas} is the global warming potential of gas.

Note: A combination of all the fuel types and GHGs have to be considered and then summed to get total emissions from the energy sector.

2. *Domestic.* This sector includes consumption in the form of cooking, heating, household appliances and lighting. The majority of the households in India uses liquefied petroleum gas (LPG) as fuel, followed by firewood, kerosene and piped natural gas (PNG). Another major consumption of energy for this sector is electricity. The pollutants and GHGs from these sources deteriorate the indoor air quality. Considering the rural population, still thousands of families rely on dried cow dung cakes as a fuel for cooking. Total emission from this sector is quantified using Equation 2.2. A few big Indian cities, for example Kolkata, have the highest GHG and emission contribution from this sector when compared to any other sector (Ramachandra et al., 2015).

3. *Transportation.* Ever since the development of the automobile sector, there has been an increased count of vehicles and the sector has been one of the biggest sources of anthropogenic GHG emissions. In this sector, road transport contributes the most emissions, and emissions from road transportation depends upon type of vehicle, type of engine, efficiency of engine, type of fuel, sophistication of technology, age of vehicle, maintenance by user, etc. Emissions from this sector can be estimated using two methods:

   a. Distance travelled by vehicles approach. This is also known as the bottom-up approach and is estimated by the number of vehicles, annual distance travelled and respective emission factor for the vehicle:

$$\text{Emission} = \text{Vehicle}_j \times D_j \times \text{Emission factor}_{j,km} \qquad (2.4)$$

   where *Vehicle*$_j$ is the number of vehicles of one type, D_j is the distance travelled by the vehicle of type j in one year, and *Emission factor*$_{j,km}$ is the emission factor of j type of vehicle per kilometre driven.

   Note: Emission here refers to one GHG only. For all the GHGs, this emission has to be estimated and summed up for total emission contributed by this sector.

   b. Consumption of fuel. This includes quantification of GHGs from the amount of fuel consumption (Klein et al., 2012). The IPCC provided guidelines in 2006 regarding the method to estimated emissions using fuel consumption as per Equation 2.2.

   Note: All vehicle types such as two wheelers (with gears and without gears), four wheelers (cars, jeeps, taxis), light motor vehicles, heavy motor vehicles, lorries, tractors, trailers, aviation, ships and cargo should be considered.

   Note: All fuel types such as gasoline (petroleum, diesel), compressed natural gas, high-speed diesel, light diesel oil and fuel oil should be taken into account.

4. *Industrial.* There are multiple and diverse sources of different GHG emissions from industrial activities. The major contribution is from industries causing chemical and physical changes in materials such as the iron and steel industry, cement industry, chemical feedstock industry, ammonium manufacturing industry, and petrochemical industry. IPCC 2006 guidelines define how calculation of emissions from each industry has to be performed. For example, emissions from ammonia production are estimated as per Equation 2.5:

$$\text{Emission} = A \times F \times \text{CCF} \times \text{COF} \times \frac{44}{12} - R_{CO_2} \qquad (2.5)$$

where A is the amount of ammonia production, F is the fuel required for per unit production, CCF is the carbon content factor, COF is the carbon oxidation factor and R_{CO_2} is the CO_2 recovered in the form of urea downstream.

Note: For every industrial product, there is variation in the estimation method.

5. *Agriculture.* In this sector, key areas that contribute to emissions are paddy cultivation, and stubble and residue burning. Methane and N_2O are the main GHGs that emit from this sector. Methane can be released into the atmosphere via three processes: rice cultivation (~90–95%), loss as a bubble from paddy soil and diffusion loss across the surface. Total methane emission can be estimated using

$$\text{Emission} = \text{EF} \times A \tag{2.6}$$

where EF is the seasonal integrated emission factor for various conditions and A is the annual harvested area for various conditions.

Nitrous oxide is produced naturally in agricultural soil due to the bacterial action in the process of nitrification and denitrification. Anthropogenic release of nitrous oxide can be divided into direct and indirect release. The main sources of direct emissions are nitrogenous rich fertilisers, organic nitrogen, urine, dung and manure, and are estimated as per Equation 2.6 (where EF is the emission factor for the source and A is amount of fertiliser, urine, dung or manure added in soil). Sources for indirect emissions include crop residue, mineralisation, and paddock by grazing animals. It is also estimated as per Equation 2.6.

6. *Waste.* Major emissions from this sector are in terms of methane and due to the disposal of municipal solid waste, domestic and industrial wastewater. As per IPCC guidelines, emission from disposal of solid waste is calculated based on first decay order and is estimated as

$$\text{Emission} = \left[\left(M \times \text{MCF} \times \text{DOC} \times \text{DOC}_f \times F \times \frac{16}{12}\right) - R\right] \times (1 - \text{OF}) \tag{2.7}$$

where M is the mass of waste, MCF is the methane correction factor, DOC is degradable organic carbon, DOC_f is the decomposable fraction of organic carbon, F is the portion of CH_4 generated in gas, R is methane recovery, 16/12 is the molecular weight of CH_4/C and OF is the oxidation factor.

Methane from wastewater is due to the presence of organic fractions and is quantified as

$$\text{Emission} = [U \times T \times \text{EF}](\text{TOW} - S) - R \tag{2.8}$$

where U is the fraction of population of organic matter, T is the degree of utilisation of the treatment pathway, TOW is the total organic in wastewater, S is the organic component removed as sludge, and R is the amount of methane recovered.

Note: Similarly, emission has to be estimated for other GHGs.

7. *Livestock.* Activities contributing to emissions from this sector include manure management and enteric fermentation in ruminants. Enteric fermentation can be defined as digestive processes by which carbohydrates break down by microbial activities to produce simple molecules for absorption into the bloodstream of ruminants. A major GHG from this sector is methane and the amount depends on the type of cattle, its age, digestive tract, quality of feed consumed, etc. Emission is estimated as per the following equation:

$$\text{Emission} = EF \times N \qquad (2.9)$$

where EF is the emission factor for the defined livestock population and N is the number of livestock species.

For emissions for manure management, EF and N are multiplied with average nitrogen excretion to determine the emissions.

8. *Fugitive.* This sector includes the deliberate or accidental emission of GHGs as a result of production, extraction, processing and transportation of fuel. Emissions from this category cannot be ignored and are estimated using

$$\text{Emission} = \text{Refinary crude throughput} \times EF \qquad (2.10)$$

Emissions from all these sectors are combined to provide net GHG release from a city.

2.5 Case Study of Delhi

GHG emissions for the city of Delhi were estimated. Delhi, being the capital city of India, has the highest annual carbon footprint in the country. It has multiple industries, has a huge transportation network and has a central station along with one of the biggest airports with high connectivity. Emissions from various sectors in Delhi were computed, and it was inferred that the highest emissions correspond to the transportation sector followed

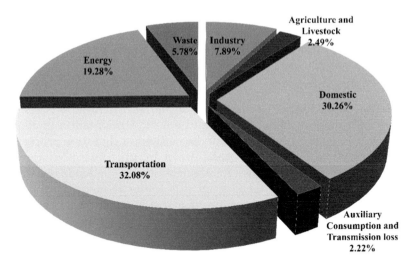

FIGURE 2.5
Sector-wise GHG emissions for Delhi.

TABLE 2.2

Sector-Wise Emissions for Delhi

Sector	Emission (CO_2e in Gg)
Energy	Commercial: 5428.55 Others*: 2099.11 Auxiliary consumption and supply losses: 857.69
Transportation	Other than CNG fuel: 10867.51 CNG fuel: 1527.03
Industrial	3049.30
Agriculture	Paddy: 17.05 Soils: 248.26 Stubble burning: 2.68
Waste	Solid waste disposal: 853.19 Domestic wastewater: 1378.75
Livestock	Enteric fermentation: 570.57 Manure management: 43.09

*Electricity consumption by streetlights, advertisement hoardings, public water works and sewerage systems, irrigation and agriculture, pumping systems, religious worship, crematoria and burial grounds

by the domestic, energy, industry, waste, agriculture, livestock and fugitive sectors (Figure 2.5). Overall emissions from Delhi were found to be 69.4 million tonnes CO_2e as per Down to Earth (2019). Table 2.2 represents sector-wise emissions for Delhi (Ramachandra et al., 2015). Per capita CO_2e was estimated to be 2.40 tonnes and per capita GDP was found to be INR 1.36 lakhs. CO_2e emissions per GDP were thus calculated and found to be 1.76.

2.6 Conclusion

India ranks second globally when it comes to demography and contributes 7% of global emissions, following the global giants – China, US and the European Union. India, being one of the fastest growing economies, has been witnessing rapid changes in land use, increased transportation, uncontrolled urbanisation, increasing demand of energy, etc., that are contributing to the enhanced GHG emissions. As a result of these issues, air quality is being sacrificed that is affecting the microclimate as well as health of residents. Delhi ranks seventh among the top polluted global cities. In this chapter, a sector-wise GHG emissions quantification framework along with details of likely implications have been communicated. A case study of the Indian megacity Delhi was provided at the end of the chapter.

References

AMS. 2012. *Climate change.* Accessed on September 26, 2018. Retrieved from https://www.ametsoc.org/ams/index.cfm/about-ams/ams-statements/statements-of-the-ams-in-force/climate-change/

Arjun, K. M. 2013. Indian agriculture- Status, importance and role in Indian economy. *International Journal of Agriculture and Food Science Technology,* 4(4), 343–346.

Berner, W., Oeschger, H., & Stauffer, B. 1980. Information on the CO_2 cycle from ice core studies. *Radiocarbon,* 22(2), 227–235.

Boucher, O., Friedlingstein, P., Collins, B., & Shine, K. P. 2009. The indirect global warming potential and global temperature change potential due to methane oxidation. *Environmental Research Letters,* 4(4), 1–5.

BSI. 2011. *Specification for the assessment of the life cycle greenhouse gas emissions of goods and services.* Accessed on March 26, 2019. Retrieved from http://shop.bsigroup.com/

Carbon Brief. October 20, 2016. *Review: 7 key scenes in Leonardo DiCaprio's climate film* Before the Flood. Accessed on May 02, 2019. Retrieved from https://www.carbonbrief.org/

CEEW. 2019. *Greenhouse gases emission estimates from the manufacturing industries in India state level estimates: 2005 to 2013.* Accessed on May 02, 2019. Retrieved from https://www.ceew.in

Central Electricity Authority. 2018. *CO_2 baseline database for the Indian power sector.* Accessed on March 28, 2019. Retrieved from http://www.cea.nic.in

Down to Earth. 2019. *India's emission capitals.* Accessed on May 03, 2019. Retrieved from https://www.downtoearth.org.in/

Fan, Y., Liu, L. C., Wu, G., Tsai, H. T., & Wei, Y. M. 2007. Changes in carbon intensity in China: Empirical findings from 1980–2003. *Ecological Economics,* 62(3–4), 683–691.

FAO. 2019. *Key facts and findings - By the numbers: GHG emissions by livestock.* Accessed on April 03, 2019. Retrieved from http://www.fao.org/

Garg, A., Kankal, B., & Shukla, P. R. 2011. Methane emissions in India: Sub-regional and sectoral trends. *Atmospheric Environment, 45*(28), 4922–4929.

GGW. 2019. *Industry CO$_2$ emissions.* Accessed on April 03, 2019. Retrieved from https ://www.global-greenhouse-warming.com/

GHG Online. 2019. *Nitrous oxide sinks.* Accessed on March 25, 2019. Retrieved from http://www.ghgonline.org/

ICLEI. 2017. *Waste sector emissions estimates for 2005–2013: National and state level.* Accessed on April 03, 2019. Retrieved from http://www.ghgplatform-india. org/

INCCA. 2010. *India: Greenhouse gas emissions 2007.* Accessed on April 04, 2019. Retrieved from http://www.moef.nic.in/

Indermühle, A., Stocker, T. F., Joos, F., Fischer, H., Smith, H. J., et al. 1999. Holocene carbon-cycle dynamics based on CO$_2$ trapped in ice at Taylor Dome, Antarctica. *Nature, 398*(6723), 121.

IPCC. 2014. *Climate change 2014 syntheses report.* Accessed on March 23, 2019. Retrieved from https://www.ipcc.ch/site/assets/uploads/2018/05/SYR_AR5_FINAL_ful l_wcover.pdf

Jayalakshmi, K. March 18, 2015. India: Transport, domestic sectors ahead in greenhouse gas emissions. *International Business Times.* Accessed on March 26, 2019. Retrieved from http://wgbis.ces.iisc.ernet.in/

Klein, J., Geilenkirchen, G., Hulskotte, J., Ligterink, N., & Molnár-in't Veld, H. 2012. Methods for calculating the emissions of transport in the Netherlands. *Task Force on Transportation of the Dutch Pollutant Release and Transfer Register.* Accessed on May 03, 2019. Retrieved from http://www.emissieregistratie.nl

Le, T. H., Somerville, R., Cubash, U., Ding, Y., Mauritzen, C., et al. 2007. Historical overview of climate change. In Solomon, S., Qin, D., Manning, M., Chen, Z., Marquis, M., et al. (Eds.), *Climate Change 2007: The physical science basis. Contribution of Working Group I to the Fourth Assessment Report of the Intergovernmental Panel on Climate Change.* Cambridge and New York: Cambridge University Press.

Marshall, M. 2009. *Timeline: Climate change.* Accessed on May 16, 2018. Retrieved from https://www.newscientist.com/article/dn9912-timeline-climate-change/

Maslin, M. 2008. *Global warming: A very short introduction.* New York: Oxford University Press.

McMichael, A. J., Woodruff, R. E., & Hales, S. 2006. Climate change and human health: Present and future risks. *The Lancet, 367*(9513), 859–869.

NASA. 2018. *Global climate change – Vital signs of the planet.* Accessed on May 15, 2018. Retrieved from https://climate.nasa.gov/scientific-consensus/

Neftel, A., Oeschger, H., Schwander, J., Stauffer, B., & Zumbrunn, R. 1982. Ice core sample measurements give atmospheric CO$_2$ content during the past 40,000 yr. *Nature, 295*(5846), 220.

NOAA. 2018. *A paleo perspective on global warming.* Accessed on September 26, 2018. Retrieved from https://www.ncdc.noaa.gov/global-warming

NOAA. 2019a. *Greenhouse gases.* Accessed on March 23, 2019. Retrieved from https://www.ncdc.noaa.gov/monitoring-references/faq/greenhouse-gases.php

NOAA. 2019b. *Climate change: Atmospheric carbon dioxide.* Accessed on March 23, 2019. Retrieved from https://www.climate.gov/news-features/understanding-cli mate/climate-change-atmospheric-carbon-dioxide

NOAA. 2019c. *Trends in atmospheric methane.* Accessed on March 23, 2019. Retrieved from https://www.esrl.noaa.gov/gmd/ccgg/trends_ch4/

Omenn, G. S., Holdren, J., Baltimore, D, Shaw, D. E., Golden, W. T., et al. 2006. *AAAS board statement on climate change.* Accessed on September 26, 2018. Retrieved from https ://www.aaas.org/sites/default/files/s3fs-public/aaas_climate_statement_0.pdf

Pachauri R.K., & Chand M. 2008. A global perspective on climate change. In Moniz, E.J. (Ed.), *Climate change and energy pathways for the Mediterranean* (Vol. 15). Alliance for global sustainability bookseries. Dordrecht: Springer.

Pappas, D., & Chalvatzis, K. J. 2017. Energy and industrial growth in India: The next emissions superpower? *Energy Procedia, 105,* 3656–3662.

Pollack, H. N., Huang, S., & Shen, P. Y. 1998. Climate change record in subsurface temperatures: A global perspective. *Science, 282*(5387), 279–281.

Ramachandra, T. V., Bharath, H. A., & Sreejith, K. 2015. GHG footprint of major cities in India. *Renewable and Sustainable Energy Reviews, 44,* 473–495.

Ramanathan, R., & Parikh, J. K. 1999. Transport sector in India: An analysis in the context of sustainable development. *Transport Policy, 6*(1), 35–45.

Ritchie, H., & Roser, M. 2019. *CO_2 and other greenhouse gas emissions.* Accessed on April 26, 2019. Retrieved from https://ourworldindata.org/

SciDev. 2018. *Methane from Indian livestock adds to global warming.* Accessed on April 03, 2019. Retrieved from https://www.scidev.net/

Sims, R., Schaeffer, R., Creutzig, F., Cruz-Núñez, X., D'agosto, M., et al. 2014. *Transport climate change 2014: Mitigation of climate change. Contribution of Working Group III to the Fifth Assessment Report of the Intergovernmental Panel on Climate Change.* Accessed on April 25, 2019. Retrieved from http://www.ipcc.ch/pdf/assessm ent-report/ar5/wg3/ipcc_wg3_ar5_chapter8.pdf

Singh, N., Mishra, T., & Banerjee, R. 2019. Greenhouse gas emissions in India's road transport sector. In Venkataraman, C., Mishra, T., Ghosh, S., & Karmakar, S. (Eds.), *Climate change signals and response* (pp. 197–209). Singapore: Springer.

Sovacool, B. K., & Brown, M. A. 2010. Twelve metropolitan carbon footprints: A preliminary comparative global assessment. *Energy Policy, 38*(9), 4856–4869.

SSEF. 2017. *GHG emissions from India's electricity sector.* Accessed on March 28, 2019. Retrieved from https://shaktifoundation.in/

Statista. 2019. *Total number of vehicles across India from 2005 to 2016 (in millions).* Accessed on April 16, 2019. Retrieved from https://www.statista.com/

TWB. 2014. *CO_2 emissions from transport (% of fuel combustion).* Accessed on April 16, 2019. Retrieved from https://data.worldbank.org/

Two Degree Institute. 2019. *Global N_2O levels.* Accessed on March 25, 2019. Retrieved from https://www.n2olevels.org/

UN Development Program. (2018). *Climate Box: An Interactive Learning Toolkit on Climate.* Accessed on March 22, 2019. Retrieved from http://www.undp.ru/i ndex.php?iso=RU&lid=1&cmd=publications1&id=176

UNCC. 2018. *The Paris agreement.* Accessed on May 16, 2018. Retrieved from https ://unfccc.int/process-and-meetings/the-paris-agreement/the-paris-agreement

UNCC. 2019. *Global warming potentials.* Accessed on March 23, 2019. Retrieved from https://unfccc.int

Wang, W. C., Yung, Y. L., Lacis, A. A., Mo, T. A., & Hansen, J. E. 1976. Greenhouse effects due to man-made perturbations of trace gases. *Science, 194*(4266), 685–690.

Wigley, T. M. L., & Raper, S. C. B. 2001. Interpretation of high projections for global-mean warming. *Science, 293*(5529), 451–454.

WYI. 2019. *Main sources of carbon dioxide emissions.* Accessed on March 23, 2019. Retrieved from https://whatsyourimpact.org/

3

Land Use and Land Cover Dynamics

Synthesis of Spatio-Temporal Patterns

CONTENTS

3.1   Introduction 39
3.2   Terminology and General Definitions 40
3.3   Global LULCC 41
3.4   Indian Scenario of LULCC 43
3.5   Causes of LULCC 44
3.6   Impacts of LULCC 45
3.7   Application of Remote Sensing and GIS to Identify LULCC 46
3.8   Case Study: Chennai Metropolitan Area 52
3.9   Concluding Remarks 55
References 56

3.1 Introduction

This chapter introduces the underlying phenomenon in landscape dynamics through the spatio-temporal analyses of land use/land cover (LULC) using spatial data acquired through space-borne sensors. The chapter begins with a brief history of anthropogenic-induced land use changes, definitions of terms, global and local perspectives, and causes and impacts of land use changes. The chapter also introduces remote sensing techniques for monitoring LULC changes through metrics such as vegetation difference indices. Understanding the landscape process and change detection requires prior knowledge of landscapes, fragmentation, ecology and ecosystem, remote sensing data analysis especially in the geospatial domain, advanced machine-learning techniques and ground truthing for validation of classification. A case study is conducted on a metropolitan city situated on the south-eastern coastline of India for better comprehension of geospatial concepts.

3.2 Terminology and General Definitions

Landscapes across the globe have witnessed changes in structure either due to natural events or human-induced activities. Transitions from towns to cities, and cities to larger agglomerations were evident during the mid-19th century with the Industrial Revolution. This period witnessed large-scale transitions from traditional agriculture or farming practices to urban pockets. This spurted growth in urban pockets in the form of sprawl with implications for infrastructure and basic amenities. The constant inflow of citizens to urban centres for livelihood and education has posed serious implications on the natural resource base necessitating local authorities to monitor the process of land acquisition and conversion to evolve appropriate strategies for sustainable management of natural resources towards the design of sustainable cities. Numerous cartographic maps have been used to assess land use changes both qualitatively and quantitatively. However, traditional mapping could not cope with the exponential increase in land use changes during the post-industrialisation period. In the early 1970s, computer-aided systems were introduced to analyse spatial data (draw and update maps). The emergence of digital technologies led to an increase in remotely sensed imaging and photographic systems for mapping changes in a landscape. Geographic information systems (GIS) with remote sensing data (space-borne satellite sensors and aerial photographs) provided new perspectives to map, analyse, monitor and update large-scale LULC changes. Four resolutions of remote sensing data are spatial, spectral, temporal and radiometric, which aid to define an area under observation to map both natural and man-made features. The features describing characteristics of Earth's surface can be broadly understood with the help of two terminologies: land cover and land use. Turner et al. (1995) and Lambin et al. (2003) define land cover as Earth's land surface and immediate subsurface (the biophysical surface) covered by natural features such as sand, soil, rock, vegetation (includes herbs, shrubs, grass, primary and secondary forest, etc.), water (includes saltwater, freshwater in the form of rivers, lakes, ponds, etc.), ice sheet and snow. Conversion of land cover would also significantly impact the surrounding ecosystem. Unless it is a human intervention, land cover conversion does not have huge impact; however, this process takes a longer time and there may be changes in biodiversity, primary productivity, oxygen and carbon cycle, soil quality, surface runoff and other attributes related directly to the surface of Earth. The natural process of land cover change takes several centuries. On the other hand, land use is defined as the land area exploited by humans for their needs. Land use reflects human interactions in a particular piece of natural land. Land use is therefore kept at a level higher than land cover on the semantic hierarchy, resulting directly in resource manipulation due to human activity or economic function associated with land type. Definition of land use might vary from region to region, depending on level of mapping detail, geographic area covered, etc. However, understanding of land use and land cover specific to a region forms the key to successfully plan and manage land reforms. Primary natural resources vulnerable to human needs include forest, grassland, wetland, etc. Changing patterns of land cover and land use over time have become an emerging concern among researchers and addressed as land use and land cover change (LULCC). The aim of this chapter is to introduce the reader to a basic understanding of LULCC. It begins with a global LULCC scenario and gradually provides causes, impacts and visualisation of LULCC using satellite-based remote sensing technologies.

3.3 Global LULCC

Humans have been altering land cover by clearing forests over the past 10,000 years. Humans started as wanderers, slowly practising shifting cultivation by expanding into forests, grasslands and steppes around all parts of the globe to meet their needs. Agricultural expansion paved way to development of civilisation, population expansion and modern economy. The last 300 years have witnessed significant changes with the predominant usage of fossil fuels, exponential population growth, rise of capitalism and advent of industrial technologies. The area of pastureland has increased from 500 million hectares to 1700–3100 million hectares in 1990 (Ramankutty et al., 2002; Lambin et al., 2003), whereas the area of natural forest cover has decreased from ca. 5000–6200 million hectares to 4300–5300 million hectares (in 1990). Eastern China, the Indo-Gangetic Plain and parts of Europe are some of the regions experiencing rapid cropland expansion, followed by other parts of the world. Right from the initial days of the 20th century, the world has seen the most rapid land cover changes in a wide spectrum of categories and some of the significant land cover changes are discussed in the following list.

1. *Forest cover changes.* Loss of forest can be due to human or natural causes. Native forest regions witnessed large-scale transformations leading to deforestation due to anthropogenic activities. Deforestation occurs as a result of converting or chopping off trees below the threshold of 10% (as per the Food and Agriculture Organization guidelines; FAO, 2016). Table 3.1 (adapted from GFRA, 2016) lists brief details of forest status around the world as of the year 2015.

2. *Agricultural and pasture area expansion.* Agricultural land area refers to land area under arable and permanent crops. Arable land includes areas occupied for growing temporary crops. Permanent crops occupy the land for long periods and need not be replanted after each harvest, such as coffee and rubber. Agricultural land use covers more than one-third of world's land area. Developing countries

TABLE 3.1

Details of Forest Status around the World

Region	Africa	Asia	Europe	North and Central America	Oceania	South America	World
Forest area (million ha, 2015)	624	593	1015	751	174	842	3999
Natural forest (million ha, 2015)	600	462	929	707	169	827	3695
Net annual forest change (million ha, 2010–2015)	–2.8	0.8	0.4	0.1	0.3	–2	–3.3

such as India and China, with very high population densities, have regulated agricultural expansion by implementing zoning. Humans have created a direct dependency on agriculture and livestock production for hundreds of years.

Historical examination of land suggests that humans aimed to maximise agricultural output by expanding land suitable for agriculture. India has approximately 1.891 million km^2 of cultivated land accounting for 57% of the total geographical area. Agriculture, apart from providing basic food and livelihood for millions in many countries, is one of the major sectors contributing to the national economy and gross domestic product. If agriculture is used extensively, it has potential to damage the natural ecosystem and degrade available natural resources, for instance, causing soil erosion, loss of soil fertility, loss of biodiversity due to excessive usage of chemicals and fertilisers to obtain better yield, etc. Growth of crops completely depends on a region, its climate, soil type and production procedure adopted. Hence, there is no uniform agricultural practice, as it depends on numerous factors such as economy, soil type, climate, rainfall, fertilisers/chemical used, region-based crop cycle and water quality.

3. *Urban cover changes.* Urban expansions are taking place rapidly with a higher rate of people migrating from rural areas to fringes and successively to core cities. In recent decades, a diversified and growing economy as well as tourism developments with the introduction of ecotourism has attracted new residents contributing to significant urban growth (Yuan & Bauer, 2007; Khwanruthai & Yuji, 2011). The existence of the first ever city in the world was around 3000 B.C. in Egypt and Mesopotamia. The numbers of cities slowly grew over time and it was evident that urbanisation gained momentum

from 600 BC to 400 AD with a few cities reaching substantial size such as Athens, Rome and Constantinople (Davis, 1955). A slow but gradual urban expansion process was observed from 1800 to 1900. The concepts of suburban areas, hinterland and rural urban fringe expansion were introduced to existing urban areas. Considering the global scenario, developing countries, still in transition, are more populated than developed countries where the transformation has been complete for two or three decades.

Researchers have tried to address LULCC globally as well as by region. Cui and Graf (2009) carried out studies on the Tibetan Plateau, primarily aiming to review land cover changes during the years 1960–2010. Studies also reveal vegetation species are highly sensitive to human activities and are easy targets to environment change (Zhang et al., 1996). Authors have concluded, with significant observations, that the tundra was replaced by scrubland in the central Tibetan Plateau, there was loss of warm and conifer forests at the south-eastern part of the plateau, and there were changes from grassland to bare land on the western border. Similarly, St Peter et al. (2018) developed methodology using spatial analysis software to create land cover maps at 1 m spatial resolution. The method involved integration of image texture metrics along with principal component analysis and neural networks. Results showed land cover variables could be evaluated with much less time using a probabilistic LC data set. Townshend et al. (2012) attempted to characterise forest cover using 30 m global Landsat data. The data set included the optimal 9500 scenes acquired around the globe during the years 2004–2007 with several compatible criteria (Gutman et al., 2008). They adopted automatic identification of a training data method to overcome paucity of high-quality data and with application of support vector machine classification to suppress errors. Global forest cover change with areas greater than 30% was reported for the first time for remote sensing data with moderate (Landsat) resolution.

3.4 Indian Scenario of LULCC

The history of LULCC and the concept of agricultural land expansion have emerged as a form of spatio-temporal discontinuities confined to the Indus Valley, Indo-Gangetic Plain and surroundings around 7000 years ago. Fertile soil, favourable climate, availability of water and sufficient land to cultivate crops has given impetus to the growth of the world's oldest civilisations (Abrol et al., 2002). During the early historical period, cities such as Pataliputra (presently Patna, Bihar state) and Varanasi in the north, Kancheepuram and Madurai in the south fuelled dynamic land use change. Craft villages were formed at the peripheries of core cities with major occupation such as agriculture, horticulture, textiles, carpentry, metal work and stonework. The next phase of rapid change in landscape was the medieval period with major business centres such as Nasik, Puri, Cuttack, Ujjain, Delhi and Meerut in the north, and Tanjore, Kumbakonam, Tiruchirapalli, Sholapur, Golconda and Hampi to name a few in the south. All these cities had unique LULCC patterns. The last phase before independence was the British period, which marked a milestone in Indian history. Studies conducted have estimated that the country lost forest cover of over 40% in a span of 100 years from 1880 (Richards & Flint, 1994). Reasons for rapid change have been identified as the increased demand for food and therefore increase in cultivated/agricultural land, increase in livestock, commercial logging, mining, hydropower and related industries among others (Sala, 1995). At the end of 1900s, Calcutta become a premier city with a population of nearly 8 lakhs and showed large-scale land cover conversions along with other British-influenced cities such as Bombay, Kanpur and Madras. By the end of British rule in India, the rail and road network had been significantly developed and covered major parts of the country, making way for permanent land cover changes. A study by Lele and Joshi (2008) focused on deforestation and accounting for spatial changes of forest cover in the north-eastern part of India during the years 1972–1999. They highlight deforestation as one of the critical components in addressing LULCC dynamics. The contribution of various agencies in monitoring forest density throughout the country has been significant. For instance, the Ministry of Agriculture & Cooperation, National Remote Sensing Agency (NRSA), Forest Survey of India, and United Nations Food and Agriculture Organization carried out national-level vegetation mapping for the entire country. With the collaborative effort of 21 institutes and 61 scientists, Roy et al. (2015) came up with a vegetation type map by studying different species, the area covered, elevation at which species were found, temperature, precipitation, etc., across the country. Other areas including North-East India (Roy & Joshi, 2002; Roy & Tomar, 2000), Western Ghats (Daniels et al., 1995; Das et al., 2006; Giriraj et al., 2008; Gadgil, 2013; Ramachandra et al., 2014) and the Himalayan region (Semwal et al., 2004; Munsi et al., 2009) have shown the importance of assessing forest land cover change periodically.

3.5 Causes of LULCC

The magnitude and scale of change on Earth's surface are unmatched, and these alterations are mostly dominated by humans. Realising the importance of this, land use and land cover changes have become the most studied topic by the scientific community since their impacts are significant and could even disturb the natural process of Earth systems (Lambin et al., 2001).

Causes of LULCC are twofold: (1) proximate causes and (2) underlying causes. Proximate causes act at the local scale and are direct in nature involving physical action on agriculture field expansion, timber logging, development of infrastructure, etc., whereas underlying causes are fundamental causes that drive more proximate causes. Proximate causes include infrastructure extension such as roads, residential and commercial establishments, natural resource extraction, agricultural land expansion, and timber extraction. Underlying causes depend on socio-economic, political, demography, tourism and other variables of a region (Brookfield, 1999).

3.6 Impacts of LULCC

Local changes in LULCC have global ramifications and hence understanding the impacts of LULCC on the ecosystem is one of the key areas of research worldwide. Developing countries such as India and China are paying the price of rapid transition in LULCC and these changes are driven mainly by unplanned developmental activities, urbanisation, population growth, etc. Causal factors of LULCC are manifold: deforestation, mining land expansion, agricultural land expansion, abandonment of agricultural fields due to lesser yield or practice of shifting agriculture, urban or pervious surface expansion, and wetland encroachment are a few among several other reasons. Imbalances in demand and supply of agricultural products, crops and food have created pressure on government agencies to draft policies with limited land available for practising agriculture. Further, the dynamics of LULCC vary by region, land management and availability, economy, and policies. However, the impacts cover a large spectrum including greenhouse gas emissions and air pollution (Chen et al., 2009), climate change (Salazar et al., 2015), alteration of biogeochemical cycles (Fu et al., 2013), hydrological cycle (Schilling et al., 2008), drought (Brian et al., 2017) local climate and environment (Bounoua et al., 2015), tree cover loss and environment quality (Bharath et al., 2018), landscape fragmentation (Taubenbock et al., 2009; Bharath et al., 2013) agricultural land loss (Pandey & Seto, 2015), surface temperature, and urban heat islands (Jiang & Tian, 2010).

Rapid LULCC degrades biodiversity and promotes imbalance in both the environment and ecosystem with the ecosystem losing the ability of sustainability and also resilience. Despite advancement in technologies, attempts are underway to find a reasonable solution while understanding nature and its setting before commencing any kind of anthropogenic activity. Shrestha et al. (2012) pointed out land fragmentation due to rapid transitions in urban land cover considering spatial and temporal land cover data through density gradient analysis. Further, patterns of fragmentation were analysed by identifying five major socio-ecological drivers: population dynamics, water provisioning, technology and transportation, institutional factors, and topography in the Phoenix Metropolitan Area (United States). The study showed that despite increased fragmentation rates in the low-density fringe, urban growth has been more contiguous than "leap frog" in pattern. During the course of change in urban land cover, native ecosystems are replaced by paved materials and buildings forcing the wipe out of trees. Loss of agricultural area and forest areas due to fragmentation is to be observed during the course of urbanisation. Pauchard et al. (2006) studied effects of urban landscape in Concepcion, a fast-growing metropolitan area in Chile, with wetlands and peri-urban forests along with rivers. These fragile ecosystems are destroyed, fragmented, degraded and invaded by non-native species. Of a total area of 32,000 hectares, net loss of wetlands accounted for 23%, and agriculture, forest and shrub land cover types for 9% during the years 1975–2000.

Recent studies have clearly demonstrated that a direct consequence of LULCC is change in the climate. Salazar et al. (2015) made an effort to estimate loss of natural vegetation land cover, being converted to other land use types with an area of 3.6 million km^2. The authors emphasised the necessity of studies in non-Amazonian regions also, which have equal potential in altering regional climate, as non-Amazonian areas have lost forest equivalent to four times that of Amazon deforestation. A total of eight biosensitive regions were taken into consideration for analysis including the Amazon Biome and Atlantic Forest among the others. They report that the conversion of wooded vegetation land cover into soybean plantations had an impact on surface temperature rise of 0.6°C, sending an alarming signal to local authorities. Many authors have conducted experiments based on climate models agreeing on variations in temperature and rainfall pattern change when forest cover is replaced by intense agriculture or plantation activities.

The aforementioned impacts not only affect the local environment but also have consequences for the globe. The degrading environment and biodiversity, deterioration of surface and ground water, and changes in vegetation cover are a few notable effects of rapid LULCC. Planners and decision makers need to focus on these key areas to minimise the negative impacts of inevitable land cover conversions. Successful planning and vision for the future growth scenario can be achieved by analysing spatial patterns of land cover changes using historical data. Remote sensing and GIS along with database management systems aid in measuring, monitoring and assessing the impacts of LULCC on the environment.

3.7 Application of Remote Sensing and GIS to Identify LULCC

Remote sensing is the science of acquiring information about objects or features from a significant distance, usually with the aid of a sensor, and not being in direct physical contact with the object or feature (Lillesand et al., 2014). Satellite remote sensing (RS) has been credited as a quick and economical technique for mapping large areas. The availability of higher spatial resolution and multi-spectral data provides the best accuracy needed for LULC mapping. Hence, Earth-observation-based monitoring of changes has been widely accepted and implemented by local, regional and national governments (Chen et al., 2000; Ji et al., 2001). Spatial data acquired through optical remote sensing based on space-borne (satellite) sensors can help map LULCC at finer scales, both temporally and spatially with consistent images of Earth's surface. The commissioning of Landsat 1 by the National Aeronautics and Space Administration (NASA) in the year 1972 provided a new avenue to monitor natural resources, thereby making it easier for policymakers and planners to understand the dynamics of landscape at greater detail (NASA, 2001). The Landsat program is credited as the longest running space-based Earth-observation satellite sensing program with a series of satellites, the latest being Landsat 8, launched in 2013. Considering the advantage of Landsat's sensing capability, it is acknowledged as one of the most reliable data sources for LULCC experiments and has been used extensively by researchers worldwide. Table 3.2 lists multi-resolution data details of Landsat missions.

TABLE 3.2

Data Details of Earth Observation Satellite – Landsat

Landsat Sensor	Spatial Resolution	Spectral Resolution	Radiometric Resolution	Temporal Resolution	Operational Status
Multi-Spectral Scanner (Landsat 1–3)	80 m	4 bands	6 bits Data range: 0–63	18 days	Decommissioned
Thematic Mapper (Landsat 4–5)	30 m	7 bands	8 bits Data range: 0–255	16 days	Decommissioned
Enhanced Thematic Mapper Plus (Landsat 7)	15 m (PAN) and 30 m	8 bands	8 bits Data range: 0–255	16 days	Currently in orbit; live satellite fuelling by 2020
Operational Land Imager and Thermal Infrared Sensor (Landsat 8)	15 m (PAN) and 30 m	11 bands	16 bits Data range: 0–65535	16 days	In orbit

Source: US Geological Survey, Landsat missions, http://www.usgs.gov/landsat.

Anderson (1976) developed a framework for land use/land cover classification using remote sensor data with an intention to provide a uniform classification system throughout the country (originally the United States). The study provided certain vital criteria to obtain an effective classified map, such as (a) minimum level of accuracy of any classified land use or land cover information not lower than 85%, (b) possibility of categories aggregation, and (c) a classification system suitable for remote sensing data of different times of a year. Researchers adopted the aforementioned framework for classifying remote sensing data to arrive at various land use and land cover classes.

Rogan and Chen (2004) conducted a review of applications of remote sensing data for mapping and monitoring LULCC, similar to Anderson (1976) and Jensen and Cowen (1999), to arrive at various levels of classification, with discussions of crucial methodological issues related to LULCC mapping such as trade-offs existing across multi-resolution remote sensing data having direct impact on classification accuracies. For instance, IKONOS data has exemplary spatial resolution of less than 1 m, and it is extremely useful in extraction of information with respect to a very limited electromagnetic spectrum (0.45–0.85 μm). However, it does not have a larger swath and therefore a narrow area of coverage compared to the wider range of spectrum covered by the Landsat series satellites (0.4–12.5 μm) and a swath of 185 km. Hence, one has to compromise on any of these resolutions depending on applications. Considering the aforementioned facts, remote sensing data, its analysis and integration with geographical GIS form a robust technology called geospatial technology to monitor and map changes in the natural land cover due to human activities. The GIS platform has entered the majority of the service sectors by providing facilities to capture, manage, store, retrieve, analyse and visualise geospatial data on a real-time basis. A further advantage of GIS comes from its distinguished built-in database, decision support system and application-specific plug-ins, making it a reliable tool for LULCC studies (Tran et al., 2015; Boori et al., 2015).

Researchers have developed indices to analyse spatial patterns in the land cover information. Differences in spectral values are used to divide satellite data into classified thematic maps. For instance, healthy vegetation has peak reflectivity (around 45% to 50%) in the near-infrared (NIR) region and low reflectivity (around 8% to 10%) in the red region of the electromagnetic spectrum as depicted in Figure 3.1. Therefore, differences between these two bands are ideal to capture vegetation dynamics and are defined by the normalised difference vegetation index (NDVI) to measure vegetation quantity and status (Justice, et al., 1986; Goward et al., 1991). NDVI is represented as

$$\text{NDVI} = \frac{\text{NIR}(\text{DN}) - \text{Red}(\text{DN})}{\text{NIR}(\text{DN}) + \text{Red}(\text{DN})} \tag{3.1}$$

NDVI values range from –1 to +1. A zero value or values closer to zero indicate built-up, soil, open area, or mining activities, whereas negative values

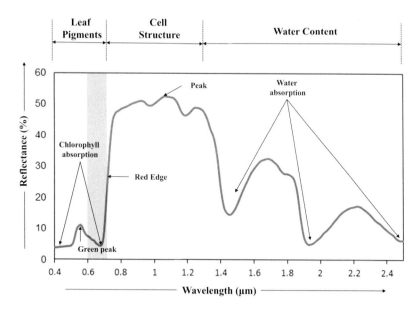

FIGURE 3.1

A typical healthy vegetation spectral reflectance curve in various regions of the electromagnetic spectrum.

indicate the presence of waterbodies. Positive values from 0.1 to 0.3 indicate sparse vegetation (shrubs, trees with some distance, etc.), 0.3 to 0.6 indicate moderate density of trees and 0.6 to 0.8 indicate dense vegetation. As the value of NDVI approaches 1, the vegetation cover becomes denser. Similar indices such as the soil adjusted vegetation index (SAVI), normalised difference water index (NDWI), normalised difference built-up index (NDBI) and normalised difference bareness index (NDBaI) have been reported by various studies (Huete, 1988; Gao, 1996; Zhao & Chen, 2005).

Bharath et al. (2017) investigated urban growth and associated land use and land cover changes in two Indian cities, namely Hyderabad and Chennai, using density gradient, spatial metrics and Shannon's entropy to quantify land cover change temporally along with NDVI analysis. To assess and visualise future changes in LC, an integrated model approach was carried out by considering two scenarios: (1) with a city development plan (CDP) and (2) without a CDP. These scenarios indicated when there is a constraint of a CDP, land use changes occur only at the periphery, away from the city boundary, essentially a leap-frog type of urban growth in contrast to the situation without a CDP wherein trends of large-scale LULCC were observed within the metropolitan boundary signifying infill-type growth. Temporal land cover results of Chennai are depicted in Figure 3.2 and corresponding statistics are given in Table 3.3.

Del Castillo et al. (2015) demonstrated the application of remote sensing techniques and GIS in monitoring and evaluating forest cover change in

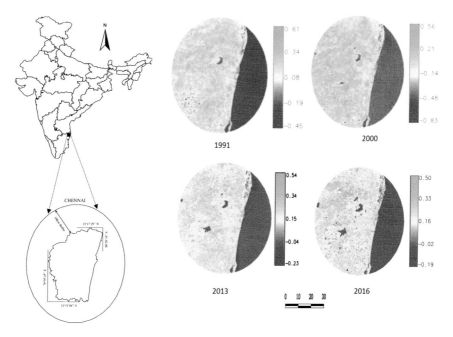

FIGURE 3.2
Temporal land cover changes of Chennai region, India, during the years 1991–2016.

TABLE 3.3

Results of Land Cover Changes for Various Cities
during the Early 1990s and Mid-2010s

City	Year	% Vegetation	% Non-Vegetation
Ahmedabad	1990	39.96	60.04
	2017	19.83	80.17
Bengaluru	1992	67	33
	2017	26.8	73.2
Chennai	1991	70.47	29.53
	2016	29.1	70.9
Delhi	1980	41.79	58.21
	2017	28.74	71.26
Hyderabad	1989	95.64	4.36
	2016	45.58	54.42
Kolkata	1990	28.73	71.27
	2017	27.66	72.34
Mumbai	1992	39.77	60.23
	2017	16.04	83.96

Moncayo Natural Park, Spain, during the years 1987–2010. A classified map was generated having nine unique classes derived from the Landsat mission series. The authors observed a decrease in pine forest and mixed transformation of mixed shrub areas with land use fragmentation.

Alqurashi and Kumar (2013), with the help of geospatial technology, accounted for land use and land cover change using the most commonly used image analysis techniques. Details of the techniques are depicted in Figure 3.3. Joshi et al. (2016) reviewed application of remote sensing and GIS to land use mapping especially focusing on image fusion techniques using optical and radar data and reported improved accuracy (in the fused image compared to single-source image analysis). Of these, pre-classification fusion was the most popularly followed algorithm supplemented by the maximum likelihood classifier. The authors categorised the literature based on (a) LULC characterisation, for e.g., broad land use, specific land use, land management; (b) study locations and study area extent; (c) various optical and radar sensors used; (d) image resolution; (e) classification method used to map LULC, for e.g., traditional, machine-learning applications, knowledge-based method; and (f) temporal frequency, for e.g., static, multi-temporal. They concluded that the majority (28) of studies infer that image fusion helps in improving the results of LULC.

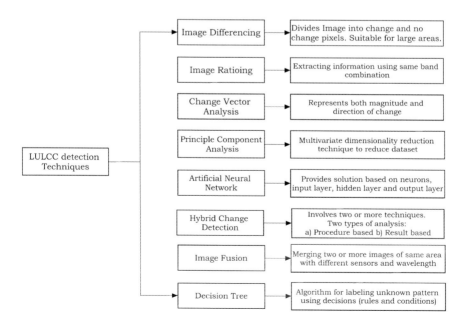

FIGURE 3.3
Techniques adopted by researchers globally to address LULCC detection temporally.

3.8 Case Study: Chennai Metropolitan Area

Considering the advantage of high spatial as well as temporal resolution, an area can be continuously monitored using satellites with sensors. Power to space-borne satellite data comes with ancillary/secondary field visit data. Field data of Chennai Metropolitan Area (CMA) was collected during May 2017 to assess land cover changes. Chennai is the capital city of Tamilnadu state, India. It is located on the eastern coast, the Coromandel Coast, also popularly known as the "Gateway to South India". The jurisdiction of the Chennai (city) Corporation was expanded from 174 km^2 to 426 km^2 in 2011. CMA has an area of 1189 km^2 comprising Chennai city district and partially extending to two districts – Kancheepuram and Tiruvallur. The study area included the Chennai administrative area with a 10 km buffer (depicted in Figure 3.2). The buffer is considered to account for spatial changes occurring in the hinterlands or peri-urban areas and visualisation of likely growth during the next decade.

The primary data (remote sensing) used in the study includes Landsat 8 (details in Table 3.2). Landsat data has a spatial resolution of 30 m, sufficient to study land cover changes, and this data can be downloaded from various public domains: US Geological Survey EarthExplorer and Glovis. Field data collection involved obtaining ground control points (GCPs) using the Global Positioning System (GPS) and capturing geotagged training data. The raw satellite data is to be geocorrected using ground control points by using secondary data along with handheld GPS data. Ground truth points of major road intersections, famous landmarks and training polygons were also collected, which aid in supervised classification/normalised difference indices for calculation of remote sensing data. The display colour assignment for any band of a multi-spectral image can be done in an entirely arbitrary manner. This process is known as generation of false colour composites (FCCs). Various band combinations can be achieved based on the user's requirement of a particular feature enquiry. In this study, we demonstrate FCCs in Figure 3.4 for a part of Chennai obtained by assigning BGR to three bands (green, red and near infrared) of a composite from Landsat 8. Creation of an FCC image directly helps in identifying heterogeneous patches in the landscape. It is important to note that vegetation appears to be red in colour in an FCC image, since reflectance of vegetation is higher in the near-infrared (NIR) band and it is assigned to the red channel. Bhuvan (bhuvan.nrsc. gov.in) and Google Earth (Earth.google.com) interfaces were used to collect data from remote areas and restricted areas where manual GPS data collection was not possible. They were also used to collect data in the form of points, lines and polygons of industries, IT companies, educational institutes, healthcare units, road networks, railway networks, natural drainage networks, restricted areas, ecologically sensitive areas, coastal zones, and so on. Geotagged images were taken to compare ground reality with satellite

FIGURE 3.4
False coloured composite for a part of Chennai obtained from Landsat 8, bands green, red and NIR. Image acquired on 23 April 2016.

image pixels in detail along with Google Earth data. Another major objective of a field visit is to observe land use/cover pattern in different categories. In this regard, interactions with local citizens, government officials and industry employees were conducted (wherever possible) to understand the region-specific pattern and changes in land use/cover occurring visibly over decades.

Figure 3.5 depicts Sholinganallur wetland and marsh lands. Change in land cover indicates that the wetlands are gradually being occupied by

1 - Shollinganallur Marsh lands; 2 - Perumbakkam toll gate;
3 - Residential areas close to lake bed

FIGURE 3.5
Synoptic view and corresponding real-time image acquired during field visit to Sholinganallur wetland and marsh lands, Chennai. (Date taken: 18 May 2017; Source: Author.)

urban cover. Sholinganallur wetland is marked under Pallikaranai marsh-land and it is the only (in Chennai city) and one of the very few surviving natural wetland ecosystems of South India. Illegal dumping of construc-tion debris, toxic solid waste and discharge of raw sewage directly into wetland area have contributed to reduction in total wetland area and there-fore resulting in death of species and replacement of native species by inva-sive species.

Figure 3.6 shows similar construction activity along the lake bed located near Tambaram–Oragadam Road. Figure 3.7 shows the rapid transition of landscape from secondary forest to urban land cover type within a 2-year time frame (2011–2013) at Periya Puliyur forest zone, situated about 60 km north-west from the core of the city. Eventually these lake beds/wetlands/

FIGURE 3.6
Construction activities on the lake bed, Tambaram–Oragadam Road, Chennai. (Date taken: 18 May 2017; Source: Author.)

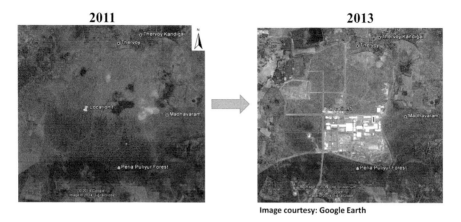

FIGURE 3.7
Periya Puliyur forest zone (outskirts of Chennai city) at two different time periods (2011–2013).
(Source: Google Earth.)

forest patches are converted into urban land use, which is a worrying sign in terms of sustainability of resources as well as the balance between humans and natural ecosystem.

3.9 Concluding Remarks

This chapter intended to provide the reader with the basic concept of LULCC with the goal of knowledge enhancement required to understand subsequent chapters of this book. In addition to fundamental understanding, causes, impacts and role of remote sensing in addressing LULCC, we also provide a case study to relate to a real-time scenario on the ground. With the advancement of remote sensing technologies, it is acknowledged that the capability of remote sensing, geoinformatics, and recent knowledge-based, artificial intelligence (machine-learning) methods collectively provide real-time global coverage with unmatched spatial and temporal scale, potentially covering a huge land surface area. Research gaps in this area such as analysis of daily monitoring of land use change, patterns, future trends and their linkage with socio-economic factors are still a challenge in developing countries due to the limitation of statistical data. Validation of LULCC still remains a challenging issue because of the dynamic nature of LC, which provide an opportunity to aspiring researchers to pursue new and integrated techniques for understanding landscape dynamics.

References

Abrol, Y. P., Sangwan, S., Dadhwal, V. K., & Tiwari, M. K. 2002. Land use/land cover in Indo-Gangetic Plains: History of changes, present concerns and future approaches. In Abrol, Y. P., Sangwan, S., & Tiwari, M. K. (Eds.), *Land use: Historical perspectives – Focus on IndoGangetic plains* (pp. 1–28). New Delhi: Allied Publishers.

Alqurashi, A. F., & Kumar, L. 2013. Investigating the use of remote sensing and GIS techniques to detect land use and land cover change: A review. *Advances in Remote Sensing*, 2(2), 193–204.

Anderson, J. R. 1976. *A land use and land cover classification system for use with remote sensor data* (Vol. 964). US Government Printing Office.

Bharath, H. A., Chandan, M. C., Vinay, S., & Ramachandra, T. V. 2017. Modelling the growth of two rapidly urbanizing Indian cities. *Journal of Geomatics*, 11(2), 149–166.

Bharath, H. A., Vinay, S., Chandan, M. C., Gouri, B. A., & Ramachandra, T. V. 2018. Green to gray: Silicon Valley of India. *Journal of Environmental Management*, 206, 1287–1295.

Bharath, H. A., Vinay, S., Durgappa, S., & Ramachandra, T. V. 2013. Modeling and simulation of urbanisation in greater Bangalore, India. In *Proceedings of National Spatial Data Infrastructure 2013 Conference*, IIT Bombay.

Boori, M. S., Vozenilek, V., & Choudhary, K. 2015. Land use/cover disturbance due to tourism in Jeseniky Mountain, Czech Republic: A remote sensing and GIS based approach. *Egyptian Journal of Remote Sensing and Space Science*, 18(1), 17–26.

Bounoua, L., Zhang, P., Mostovoy, G., Thome, K., Masek, J., et al. 2015. Impact of urbanization on US surface climate. *Environmental Research Letters*, 10(8), 084010.

Brian, F. T., James, S. F., Felix, W. L., David, N. W., Noah, P. M., et al. 2017. GRACE groundwater drought index: Evaluation of California Central Valley ground-water drought. *Remote Sensing of Environment*, 198, 384–392.

Brookfield, H. 1999. Environmental damage: Distinguishing human from geophysical causes. *Environ Hazards*, 1, 3–11.

Chen, J., Avise, J., Guenther, A., Wiedinmyer, C., Salathe, E., et al. 2009. Future land use and land cover influences on regional biogenic emissions and air quality in the United States. *Atmospheric Environment*, 43(36), 5771–5780.

Chen, S., Zheng, S., & Xie, C. 2000. Remote sensing and GIS for urban growth in China. *Photogrammetric Engineering and Remote Sensing*, 66(10), 593–598.

Cui, X., & Graf, H.-F. 2009. Recent land cover changes on the Tibetan Plateau: A review. *Climatic Change*, 94(1–2), 47–61.

Daniels, R. J. R., Gadgil, M., & Joshi, N. V. 1995. Impact of human extraction on tropical humid forests in the Western Ghats Uttara Kannada, South India. *The Journal of Applied Ecology*, 32(4), 866.

Das, A., Krishnaswamy, J., Bawa, K. S., Kiran, M. C., Srinivas, V., Kumar, N. S., & Karanth, K. U. 2006. Prioritisation of conservation areas in the Western Ghats, India. *Biological Conservation, 133*(1), 16–31.

Davis, K. 1955. The origin and growth of urbanization in the world. *American Journal of Sociology*, 60(5), 429–437.

Del Castillo, E. M., Garcia-Martin, A., Aladren, L. A. L., & de Luis, M. 2015. Evaluation of forest cover change using remote sensing techniques and landscape metrics in Moncayo Natural Park (Spain). *Applied Geography*, 62, 247–255.

Food and Agriculture Organization (FAO). 2016. *State of the world's forests 2016. Forests and agriculture: Land-use challenges and opportunities*. Rome: Food and Agriculture Organization of the United Nations.

Fu, Y., Lu, X., Zhao, Y., Zeng, X., & Xia, L. 2013. Assessment impacts of weather and land use/land cover (LULC) change on urban vegetation net primary productivity (NPP): A case study in Guangzhou, China. *Remote Sensing*, 5(8), 4125–4144.

Gadgil, M. 2013. Western Ghats: A lifescape. *Journal of the Indian Institute of Science*, 76(4), 495.

Gao, B. 1996. NDWI—A normalized difference water index for remote sensing of vegetation liquid water from space. *Remote Sensing of Environment*, 58(3), 257–266.

Giriraj, A., Irfan-Ullah, M., Murthy, M. S. R., & Beierkuhnlein, C. 2008. Modelling spatial and temporal forest cover change patterns (1973-2020): A case study from South Western Ghats (India). *Sensors*, 8(10), 6132–6153.

Global Forest Resources Assessment (GFRA). 2016. *How are the world's forests changing?* Rome: Food and Agriculture Organization of the United Nations.

Goward, S. N., Markham, B., Dye, D. G., Dulaney, W., & Yang, J. 1991. Normalized difference vegetation index measurements from the advanced very high resolution radiometer. *Remote Sensing of Environment*, 35(2–3), 257–277.

Gutman, G., Byrnes, R. A., Masek, J., Covington, S., Justice, C., et al. 2008. Towards monitoring land-cover and land-use changes at a global scale: The Global Land Survey 2005. *Photogrammetric Engineering and Remote Sensing*, 74(1), 6–10.

Huete, A. R. 1988. A soil-adjusted vegetation index (SAVI). *Remote Sensing of Environment*, 25(3), 295–309.

Jensen, J. R., & Cowen, D. C. 1999. Remote sensing of urban/suburban infrastructure and socio-economic attributes. *Photogrammetric Engineering and Remote Sensing*, 65, 611–622.

Ji, C., Liu, Q., Sun, D., Wang, S., Lin, P., et al. 2001. Monitoring urban expansion with remote sensing in China. *International Journal of Remote Sensing*, 22(8), 1441–1455.

Jiang, J., & Tian, G. 2010. Analysis of the impact of land use/land cover change on land surface temperature with remote sensing. *Procedia Environmental Sciences*, 2, 571–575.

Joshi, N., Baumann, M., Ehammer, A., Fensholt, R., Grogan, K., et al. 2016. A review of the application of optical and radar remote sensing data fusion to land use mapping and monitoring. *Remote Sensing*, 8(1), 70.

Justice, C. O., Holben, B. N., & Gwynne, M. D. 1986. Monitoring East African vegetation using AVHRR data. *International Journal of Remote Sensing*, 7, 1453–1474.

Khwanruthai, B., & Yuji, M. 2011. Site suitability evaluation for ecotourism using GIS & AHP: A case study of Surat Thani Province, Thailand. *Procedia Social and Behavioral Sciences*, 21, 269–278.

Lambin, E. F., Geist, H. J., & Lepers, E. 2003. Dynamics of land-use and land-cover change in tropical regions. *Annual Review of Environment and Resources*, 28(1), 205–241.

Lambin, E. F., Turner, B. L., Geist, H. J., Agbola, S. B., Angelsen, A., et al. 2001. The causes of land-use and land-cover change: Moving beyond the myths. *Global Environmental Change*, 11(4), 261–269.

Lele, N., & Joshi, P. K. 2008. Analyzing deforestation rates, spatial forest cover changes and identifying critical areas of forest cover changes in North-East India during 1972–1999. *Environmental Monitoring and Assessment*, 156(1–4), 159–170.

Lillesand, T.M., Kiefer, R.W., & Chipman, J. W. 2014. *Remote sensing and image interpretation*. Chichester: Wiley.

Munsi, M., Malaviya, S., Oinam, G., & Joshi, P. K. 2009. A landscape approach for quantifying land-use and land-cover change (1976–2006) in middle Himalaya. *Regional Environmental Change*, *10*(2), 145–155.

NASA/Goddard Space Flight Center—EOS Project Science Office. 2001. New satellite maps provide planners improved urban insight. *ScienceDaily*. Retrieved December 16, 2018 from http://www.sciencedaily.com/

Pandey, B., & Seto, K. C. 2015. Urbanization and agricultural land loss in India: Comparing satellite estimates with census data. *Journal of Environmental Management*, *148*, 53–66.

Pauchard, A., Aguayo, M., Pena, E., & Urrutia, R. 2006. Multiple effects of urbanization on the biodiversity of developing countries: The case of a fast-growing metropolitan area (Concepción, Chile). *Biological Conservation*, *127*(3), 272–281.

Ramachandra, T. V., Bharath, S., & Bharath, H. A. 2014. Spatio-temporal dynamics along the terrain gradient of diverse landscape. *Journal of Environmental Engineering and Landscape Management*, *22*(1), 50–63.

Ramankutty, N., Foley, J. A., & Olejniczak, N. J. 2002. People on the land: Changes in global population and croplands during the 20th century. *Ambio*, *31*(3), 251–257.

Richards, J., & Flint, E. 1994. *Historic land use and carbon estimates for South and Southeast Asia: 1880–1980*. Oak Ridge: Environmental System Science Data Infrastructure for a Virtual Ecosystem; Carbon Dioxide Information Analysis Center, Oak Ridge National Laboratory.

Rogan, J., & Chen, D. 2004. Remote sensing technology for mapping and monitoring land-cover and land-use change. *Progress in Planning*, *61*(4), 301–325.

Roy, P. S., & Joshi, P. K. 2002. Forest cover assessment in north-east India – the potential of temporal wide swath satellite sensor data (IRS-1C WiFS). *International Journal of Remote Sensing*, *23*(22), 4881–4896.

Roy, P. S., & Tomar, S. 2000. Biodiversity characterization at landscape level using geospatial modelling technique. *Biological Conservation*, *95*(1), 95–109.

Roy, P. S., Behera, M. D., Murthy, M. S. R., Roy, A., Singh, S., et al. 2015. New vegetation type map of India prepared using satellite remote sensing: Comparison with global vegetation maps and utilities. *International Journal of Applied Earth Observation and Geoinformation*, *39*, 142–159.

Sala, O.E. 1995. Human-induced perturbations, biodiversity and ecosystem functioning. In Mooney, H.A., Lubchenco, J., Dirzo, R., & Sala, O.E. (Eds.), *Global biodiversity assessment* (pp. 318–323). Cambridge: Cambridge University Press.

Salazar, A., Baldi, G., Hirota, M., Syktus, J., & McAlpine, C. 2015. Land use and land cover change impacts on the regional climate of non-Amazonian South America: A review. *Global and Planetary Change*, *128*, 103–119.

Schilling, K. E., Jha, M. K., Zhang, Y. K., Gassman, P. W., & Wolter, C. F. 2008. Impact of land use and land cover change on the water balance of a large agricultural watershed: Historical effects and future directions. *Water Resources Research*, *44*(7), W00A09.

Semwal, R., Nautiyal, S., Sen, K. K., Rana, U., Maikhuri, R. K., et al. 2004. Patterns and ecological implications of agricultural land-use changes: A case study from central Himalaya, India. *Agriculture, Ecosystems & Environment*, *102*(1), 81–92.

Shrestha, M. K., York, A. M., Boone, C. G., & Zhang, S. 2012. Land fragmentation due to rapid urbanization in the Phoenix Metropolitan Area: Analyzing the spatio-temporal patterns and drivers. *Applied Geography*, 32(2), 522–531.

St Peter, J., Hogland, J., Anderson, N., Drake, J., & Medley, P. 2018. Fine resolution probabilistic land cover classification of landscapes in the south-eastern United States. *ISPRS International Journal of Geo-Information*, 7(3), 107.

Taubenböck, H., Wegmann, M., Roth, A., Mehl, H., & Dech, S. 2009. Analysis of urban sprawl at mega city Cairo, Egypt using multisensoral remote sensing data, landscape metrics and gradient analysis. In *Proceedings of the ISRSE conference, Stresa, Italy. S* (Vol. 4).

Townshend, J. R., Masek, J. G., Huang, C., Vermote, E. F., Gao, F., et al. 2012. Global characterization and monitoring of forest cover using Landsat data: Opportunities and challenges. *International Journal of Digital Earth*, 5(5), 373–397.

Tran, H., Tran, T., & Kervyn, M. 2015. Dynamics of land cover/land use changes in the Mekong Delta, 1973–2011: A remote sensing analysis of the Tran Van Thoi District, Ca Mau Province, Vietnam. *Remote Sensing*, 7(3), 2899–2925.

Turner, B. L., Skole, D., Sanderson, S., Fischer, G., Fresco, L., et al. 1995. Land-use and land-cover change: Science/research plan. In *IGBP global change report 35/HDP rep. 7*. Stockholm: International Geosphere-Biosphere Programme.

Yuan, F., & Bauer, E. 2007. Comparison of impervious surface area and normalized difference vegetation index as indicators of surface urban heat island effects in Landsat imagery. *Remote Sensing of Environment*, 106, 375–386.

Zhang, X., Yang, D., Zhou, G., Liu, C., & Zhang, J. 1996. Model expectation of impacts of global climate change on biomes of the Tibetan Plateau. In Omasa, K., Kai, K., Taoda, H., Uchijima, Z., Yoshino, M. (Eds.), *Climate change and plants in East Asia* (pp. 25–38). Tokyo: Springer.

Zhao, H., & Chen, X. 2005. Use of normalized difference bareness index in quickly mapping bare areas from TM/ETM+. In *International Geoscience and Remote Sensing Symposium*, 3(25–29), 1666–1668.

4

Spatial Metrics

Tool for Understanding Spatial Patterns of
Land Use and Land Cover Dynamics

CONTENTS

4.1   Introduction 61
4.2   Emergence of Spatial Metrics 62
4.3   Spatial Metrics: Introduction and Applications 62
4.4   Studies Based on Spatial Metrics 63
4.5   Case Study: Hyderabad 64
    4.5.1   Data 64
    4.5.2   Methods 65
    4.5.3   Remote Sensing Data Classification and Accuracy
           Assessment 65
    4.5.4   Assessment of Landscape Dynamics through Spatial
           Metric Analysis 68
    4.5.5   Zonal Analysis 69
    4.5.6   Gradient Analysis 69
    4.5.7   Results 69
4.6   Conclusion 76
References 77

4.1 Introduction

Landscape is a mosaic of dynamic heterogeneous interacting elements. Changes in land cover would alter the structure of a landscape affecting the functional aspects of an ecosystem. Hence, understanding landscape dynamics would help in evolving appropriate strategies to mitigate the drastic changes. This chapter introduces the metrics that help in understanding landscape dynamics, which aid in understanding the spatial arrangement of natural resources (green/open spaces, lakes), distribution and spatial linkages, essential for prioritising/zoning the regions based on conservation importance as well as ecosystem services. Spatial metrics capture the status of a landscape with changing patterns, especially useful in identifying ecologically fragile regions, design of sustainable urban and also in the landscape modelling.

4.2 Emergence of Spatial Metrics

Forman and Godron (1986) defined landscape as a heterogeneous land area composed of a cluster of interacting ecosystems which has reappearance of the land forms in the entire geographical area. Landscape metrics were initially used in ecological studies (Li et al., 2005) to describe ecologically explicit relationships such as connectivity and adjacency of habitat reservoirs (Jim & Chen, 2009). Now, landscape metrics, also known as spatial metrics, along with geospatial technologies have been useful in understanding land use land cover (LULC) dynamics. Now, deforestation has been acknowledged as the prime driver of changes in the climate. This has necessitated understanding and quantification of land use/land cover changes (LULCC) and patterns (Kong et al., 2012). The landscape metrics are classified based on spatial characteristics into two types: (1) metrics that quantify the composition of the map without any spatial attributes linked to it and (2) metrics quantifying spatial configuration in the spatial data, which has spatial information as base data for the calculations (McGarigal & Marks, 1995; Gustafson, 1998). This chapter demonstrates the usage of spatial metrics to monitor spatial patterns of land use dynamics, especially to examine urban dynamics (Taubenbock et al., 2009).

4.3 Spatial Metrics: Introduction and Applications

Landscape metrics are broadly divided into six types: (1) area and edge metrics, (2) shape metrics, (3) core-area metrics, (4) contrast metrics, (5) aggregation metrics and (6) diversity metrics. These metrics are further divided into subcategories involving parameters or indices, such as edge density index, number of patches, landscape shape index and patch density. Computation of all metrics results in redundancy, and prioritised subcategory metrics would provide significant and unique information (Schneider et al., 2005). However, the selection of spatial metrics depends upon the study region (Furberg & Ban, 2012) as well as knowledge of previous studies (Hepinstall-Cymerman et al., 2013; Wu, 2006). Spatial metrics considered with their description will be discussed in the methods section. Spatial metric approach requires extensive data such as population, GDP, land use maps, floor-area ratio, maps of roadways/highways and urban city centre spatial maps (Ramachandra et al., 2012). Density gradient accompanied by time series spatial data analysis with metrics helps in the favourable output visualisation and understanding of urban dynamics (Torrens & Alberti, 2000). Thomas (1981) considered population data as primary indicator to measure urban sprawl. Studies show the usage of Shannon's entropy model to identify degree of spatial distribution of land use types and also its application of differentiating types of sprawl (Hurd et al., 2001). Many research articles demonstrate the effective integration of spatial metrics along with geoinformatics (GIS, remote sensing data and/or statistical analysis) provided insights to sprawl and its dynamics (Herold et al., 2003; Sun et al., 2013; Galster et al., 2001).

4.4 Studies Based on Spatial Metrics

O'Neill et al. (1988) studied three metrics, i.e., dominance, contagion and shape, followed by Turner et al. (1989) who considered patch size and perimeter, patch type proportion, perimeter fractal dimension, simple edge contrast and patch adjacency using FORTRAN programming language and SPAN (spatial analysis). Later, McGarigal and Marks (1995) provided a brief guideline to use a spatial pattern analysis program for quantifying landscape structures popularly known as FRAGSTATS, providing spatial statistics and metrics at three levels: (1) patch, (2) class and (3) landscape. Numerous studies have classified spatial metrics and their usage based on various factors. For instance, Thapa and Murayama (2009) concentrated on the urban development at Kathmandu Valley in Nepal. The authors aimed to analyse spatio-temporal patterns of this region with the help of satellite data (from the years 1967 to 2007) and spatial metrics. They employed eight different metrics, including number of patches, edge density and largest patch index, to assess land use fragmentation. Similarly, Galster et al. (2001) showed eight conceptually distinct, objective dimensions of land use patterns like continuity, centrality and concentration for characterising sprawl. Spatial metrics used for urban studies have been classified into three categories by Tsai (2005): density, diversity and spatial structure pattern. More important, Angel et al. (2007) used five spatial metrics with attributes and different metrics under each attribute for identifying and quantifying sprawl. Various studies (Herold et al., 2003; Wu 2006) suggested that authors have incorporated spatial metrics to measure and characterise rapid land use changes to ensure landscape planning and management (Leitao & Ahern, 2002), which remain key applications apart from detection of changes in vegetation and water cover, assessing impacts of urbanisation on landscape along with GIS (Bhatta et al., 2010; Sudhira et al., 2004).

4.5 Case Study: Hyderabad

Hyderabad is a capital of Telangana state and Andhra Pradesh (after partition in 2014). The city is located along the banks of the Musi River and surrounded by many lakes including Osman Sagar, Himayat Sagar and Hussain Sagar (Figure 4.1). Hyderabad is one of the major contributors of the national gross domestic product. With creation of special economic zones (SEZs) at Gachibowli, Pocharam and Manikonda to encourage national as well as international industries and institutions to set up operations. Erstwhile, the Hyderabad Urban Development Authority (HUDA) was expanded in 2008 to the Hyderabad metropolitan area, monitored by the Hyderabad Metropolitan Development Authority (HMDA) covering 7257 km² and a population of 7.74 million (2011). HMDA is spread across five districts, namely Hyderabad, Rangareddy, Medak, Mehaboobnagar and Nalgonda. Hyderabad is at the verge of attaining megacity status (urban agglomerations greater than ten million inhabitants), and India already has three megacities, namely Mumbai, Delhi and Kolkata (UNDESA, 2011).

4.5.1 Data

Historical remote sensing data (Landsat satellite series images) were collected from the public repository of the US Geological Survey (USGS) for three decades (1989–2017). These data were geocorrected and pre-processed to maintain spatial integrity without any distortions. The administrative boundary of Hyderabad was extracted from the city development plan (CDP),

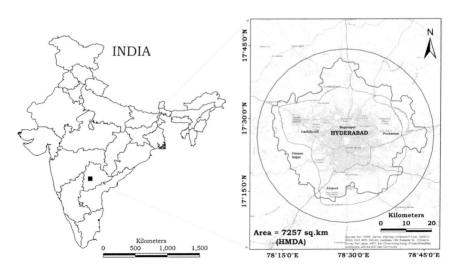

FIGURE 4.1
Study area: Hyderabad and surroundings.

and topographic maps were obtained from the Survey of India online digital repository. The study was carried out for the region covering the Hyderabad administrative boundary with 10 km buffer to understand land use dynamics in peri-urban regions at the city outskirts.

4.5.2 Methods

Figure 4.2 depicts workflow adopted for the analyses of spatial patterns of landscape dynamics within the study area. The methods included data preparation, land use classification, accuracy assessment, zonal analysis, gradient analysis and computation of spatial metrics.

4.5.3 Remote Sensing Data Classification and Accuracy Assessment

Significant spatial information is extracted from the remote sensing data, when it is processed properly. Georeferencing the data becomes the preliminary exercise before carrying out further analysis. It can be defined as aligning geographic data to a known coordinate system so it can be viewed, queried and analysed with other geographic data. Classifying corrected data helps to visualise different land use categories. The classification system for remote sensing data has to meet specific criteria as defined by Anderson (1971), for instance the classification system should be applicable over extensive areas, aggregation of categories must be possible and comparison with future land use data should be possible. There are various levels in the classification, which form a hierarchical structure.

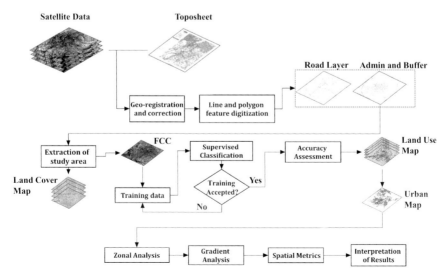

FIGURE 4.2
Method adopted to quantify temporal land use change in Hyderabad region.

Remote sensing data (digital image) classification has been a prominent approach to recognise various land use categories in a region. In this process each pixel is assigned a class based on its reflectance value to arrive at thematic maps, which give better understanding of the real-world land use types such as built-up, vegetation and water. There are different perspectives in the classification process, and the process itself tends to be subjective, even when an objective numerical approach is used. Therefore, each classification system is based on user-defined objectives. Spectrally oriented classification involves two major types: (1) supervised classification and (2) unsupervised classification. In the supervised classification schema, the image analyst supervises the pixel categorisation by specifying numerical descriptors of various land cover types present in the scene. This method uses the independent information from spectral reflectance defining training data to determine classification categories.

A classification is not complete until its accuracy is assessed. Congalton and Green (2008) have shown that classification accuracy can be obtained through error matrix or confusion matrix. The relationship between reference data and classified information is obtained by comparing them pixel-wise in the form of an error matrix. Basic terms of an error matrix are explained next.

- Rows: Represent pixels that are actually classified in each category by the classifier.
- Columns: Represent pixels that correspond to ground truth data or training data.
- Omission: Pixels that should have actually been classified according to true categories but were omitted from that category.
- Commission: Pixels that do not belong to a particular category but those which were improperly included to that category.
- Producer accuracy: It is obtained by dividing the number of correctly classified pixels in each category by the number of training set pixels used for that category (column total). These statistics indicate the reliability of training set pixels for each category of classified image. Producer accuracy corresponds to error of omission.
- User accuracy: It can be computed by dividing the number of correctly classified pixels in each category by the total number of pixels that were classified in that category (row total). User accuracy defines the probability that a pixel classified into a given category representing the same category on the ground. It corresponds to error of commission.
- Overall accuracy: It is the ratio of number of correctly classified pixels in each category to total number of pixels.
- Kappa or KHAT: It is the measure of the difference between the actual agreement between reference data and an automated classifier

TABLE 4.1

Land Use Classes Considered for Analysis

Land Use Class	Features Included in the Class
Urban	Residential areas, industrial areas, all paved surfaces, mixed pixels covering major part as built up
Vegetation	Secondary and tertiary forest, nurseries, agriculture crops greater than 2 m height and plantations
Water	Lakes, reservoirs, rivers and oceans
Others	Mining area, rocks, agricultural fields, open area, barren land

and the chance agreement between the reference data and a random classifier. Kappa coefficient can be computed using the following equation:

$$k = \frac{\text{observed accuracy} - \text{chance agreement}}{1 - \text{chance agreement}} \qquad (4.1)$$

The Gaussian maximum likelihood supervised classifier (GMLC) was employed to perform land use analysis by considering four major categories listed in Table 4.1. Land use details of selected cities of India are summarised in Table 4.2, which highlight a considerable increase in the urban areas, reduction of green cover, waterbodies and an increase in the 'Others' category due to conversion of primary/secondary forests, plantations, etc. into open lands.

TABLE 4.2

Statistics of Land Use Change for Various Cities during the Early 1990s and Mid-2010s

City	Year	Urban	Vegetation	Water	Others	OA	K
Ahmedabad	1990	7.03	36.63	2.33	54.01	92	0.84
	2017	21.36	19.74	0.72	58.18	93	0.87
Bengaluru	1992	5.47	17.01	2.42	75.1	94	0.87
	2017	24.53	5.79	0.7	68.98	90	0.78
Chennai	1991	1.46	1.38	27.64	69.52	92	0.92
	2016	22	1.83	28.34	47.83	87	0.81
Delhi	1980	9.71	38.22	0.9	51.17	99	0.99
	2017	32.37	22.03	1.01	44.59	84	0.91
Hyderabad	**1989**	**1.75**	**4**	**3.75**	**90.5**	**94**	**0.73**
	2016	**24.18**	**2.43**	**0.64**	**72.75**	**95**	**0.88**
Kolkata	1990	4.12	23	4.19	68.7	88	0.94
	2017	11.58	22.18	6.24	60	91	0.89
Mumbai	1992	7.37	21.92	45.82	24.89	78	0.81
	2017	24.14	10.02	43.76	22.08	92	0.89

Note: OA: Overall accuracy; K: Kappa.

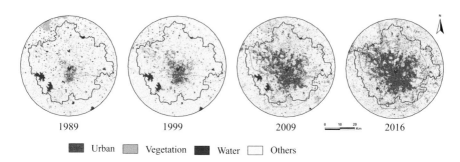

FIGURE 4.3
Spatio-temporal land use maps of Hyderabad region (1989–2016).

Figure 4.3 represents land use dynamics for Hyderabad during the past three decades with significant changes in all categories. Results reveal the steep increase in built-up areas of 93% (1989–1999), 319% (1999–2009) and 70% (2009–2016). It is important to notice that urban areas show significant increase during the years 2000–2010. This can be attributed with emergence of various industrial bases in the automobile and hardware manufacturing as well as information technology parks directly affecting the surrounding land use. The Others category has consistently reduced from 90.5% to 72.75% (1989–2016) indicating a large scale conversion of agricultural lands or bare soil land use types to urban. Waterbodies show a very critical decrease specifying either these land uses are converted or they have dried up. Statistics show a drop in the values from 3.75% to 0.64% during the past three decades suggesting an immediate intervention for rejuvenation of lakes.

4.5.4 Assessment of Landscape Dynamics through Spatial Metric Analysis

Spatial metrics are the measurements derived from the digital analysis of thematic categorical maps exhibiting spatial heterogeneity of patches, classes of patches or entire landscape mosaics of a geographic area (O'Neill et al., 1988; Herold et al., 2005). These metrics give quantitative descriptions on the composition and configuration of the urban landscape. Urban spatial metrics were drawn for the four major zonal divisions quantifying sprawl within each concentric circle of additive radius and their centre coinciding with the centroid of the study area. Details of the centroid location are given in Table 4.3. Urban sprawl pattern is understood with the help of eight different metrics (listed in Table 4.4), which were selected based on the literature (Thapa & Murayama, 2009; Herold et al., 2003; Bharath & Ramachandra, 2013), and calculations were made zone-wise for each circle of a kilometre radius. These metrics showed characteristics such as shape and contagion. The case of Hyderabad has been considered and results are discussed (Figure 4.4).

TABLE 4.3

Centroid and Concentric Circles

Centroid Latitude and Longitude		Area	Total Number of Concentric Circles
17°27'8"	78°28'56"	Begumpet (refer to Figure 4.1 and Figure 4.4)	33

4.5.5 Zonal Analysis

Most researchers have studied urbanisation patterns based on political boundaries (Taubenbock et al., 2009; Jain et al., 2011). However, to understand the growth locally specific to regions and neighbourhood, the entire study area can be divided into zones and cardinal directions (Ramachandra et al., 2012), i.e., North East, North West, South East and South West. These four zones help in identifying casual factors and degree of urbanisation at finer levels and visualising different forms of urban sprawl. Historic spatial growth in terms of direction aids in developing additional transport infrastructure connecting to the core city, identifying key economic zones and the future impact on open land, and understanding fragmentation at the periphery of the city.

4.5.6 Gradient Analysis

In addition to zone division, each zone can be divided into concentric circles with respect to the centroid of the central business district taking an increment of 1 km. These finely divided zones and concentric circle approach also help to interpret, quantify and visualise forms of urban sprawl pattern (low density, ribbon or leaf-frog development) and agents responsible for the same at local levels spatially along with classified data (Ramachandra et al., 2014).

4.5.7 Results

1. Number of patches (NP) indicate the level of fragmentation in a built-up landscape. NP also gives the count of urban or built up patches. Figure 4.5 shows NP for Hyderabad. Patches have increased in all periods of time, but years 2009 and 2017 show rapid growth in all directions representing fragmentation in these years appear to be high. It is observed that in 2017, in the core city area (circles 1–10), each patch has agglomerated into a single large urban patch, i.e., there is a saturated urban landscape with no other categories. More NPs in the NE, NW and SW directions highlight sprawl.

2. The normalised landscape shape index (NLSI) provides measure of class aggregation. All four zones show lesser values of NLSI in 2017 compared to 1989 as seen in Figure 4.6. These minimum values

TABLE 4.4

Spatial Metrics Used for Analysis and Their Significance

Indicator	Metric Type and Formula	Range	Significance/Description
Number of patches (NP)	NP = n (number of patches in landscape)	NP > 0, without limit	Measures extent of fragmentation. Higher NP, higher fragmentation.
Normalised landscape shape index (NLSI)	$$NLSI = \frac{e_i - \min e_i}{\max e_i - \min e_i}$$ where e_i = total length of edge (or perimeter) of class i in terms of number of cell surfaces; includes all landscape boundary and background edge segments involving class i $\min e_i$ = minimum total length of edge $\max e_i$ = maximum total length of edge	$0 \leq NLSI \leq 1$	nLSI = 0 when the landscape consists of a single square or maximally compact (i.e., almost square) patch. LSI increases as the patch type becomes increasingly disaggregated and reaches 1 when the patch type is maximally disaggregated.
Clumpiness (CLUMPY)	$\dfrac{G_i - P_i}{P_i}$ for $G_i < P_i$ and $P_i < 5$; else $\dfrac{G_i - P_i}{1 - P_i}$ where $$G_i = \frac{g_{ii}}{\left(\sum_{k=1}^{m} g_{ik} \right) - \min e_i}$$ g_{ii} = number of like adjacencies between pixels of patch type (class) i based on double count method g_{ik} = number of like adjacencies between pixels of patch type (class) i and k based on double count method min ei = minimum perimeter of patch type (class) i for maximally clumped class Pi = proportion of the landscape occupied by patch type (class) i	$-1 \leq$ CLUMPY ≤ 1	Shows frequency with which different pairs of patch types appear side by side. CLUMPY = -1 when the patch type is maximally disaggregated. CLUMPY = 0 when the patch type is distributed randomly, and approaches 1 when the patch type is maximally aggregated.

(Continued)

TABLE 4.4 (CONTINUED)

Spatial Metrics Used for Analysis and Their Significance

Indicator	Metric Type and Formula	Range	Significance/Description
Landscape shape index (LSI)	$$\text{LSI} = \frac{e_i}{\min e_i}$$ where e_i = total length of edge (or perimeter) of class i in terms of number of cell surfaces; includes all landscape boundaries and background edge segments involving class i $\min e_i$ = minimum total length of edge (or perimeter) of class i in terms of number of cell surfaces	LSI ≥ 1, without limit	LSI provides a simple measure of patch aggregation or disaggregation.
Patch density (PD)	$$\text{PD} = \frac{N}{A} * (10000 * 100)$$ where N = total number of patches in the landscape A = total landscape area (m²)	PD > 0, constrained by cell size	PD expresses number of patches on a per unit area basis that facilitates comparisons among landscapes of varying size.
Aggregation index (AI)	$$\text{AI} = \left[\sum_{i=1}^{m}\left(\frac{g_{ii}}{\max \to g_{ii}}\right)P_i\right](100)$$ g_{ii} = number of like adjacencies between pixels of patch type P_i = proportion of landscape comprised of patch type	0 ≤ AI ≤ 100	AI is calculated from an adjacency matrix which shows the frequency with which different pairs of patch types (including like adjacencies between the same patch type) appear side by side on the map.
Largest patch index (LPI)	$$\text{LPI} = \frac{\max a_{ij}}{A} * (100)$$ where a_{ij} = area (m²) of patch ij A = total landscape area (m²)	0 < LPI ≤ 100	LPI quantifies the percentage of total landscape area comprised by the largest patch.
Interspersion and juxtaposition (IJI)	$$\text{IJI} = \frac{-\displaystyle\sum_{i=1}^{m}\sum_{k=i+1}^{m}\left[\left(\frac{e_{ik}}{E}\right)\cdot\ln\left(\frac{e_{ik}}{E}\right)\right]}{\ln\left(0.5\left[m(m-1)\right]\right)}(100)$$ where e_{ik} = total length (m) of edge in landscape between patch types (classes) i and k E = total length (m) of edge in landscape, excluding background m = number of patch types (classes) present in the landscape, including the landscape border	0 ≤ IJI ≤ 100	IJI approaches 0 when the corresponding patch type is adjacent to only 1 other patch type and the number of patch types increases. IJI = 100 when the corresponding patch type is equally adjacent to all other patch types

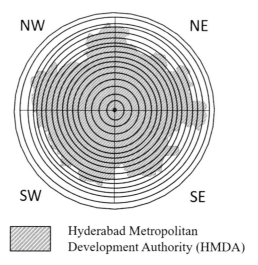

FIGURE 4.4
Graphical representation of zones and gradient for Hyderabad region. (Note: Figure is for representation purposes only).

(NLSI < 0.5) points out that the landscape consists of a single square urban patch or it is maximally compact (i.e., almost square) in contrast with the higher values in 1989 (NLSI ≈ 1) indicating that the urban patches are disaggregated maximally with complex shapes.

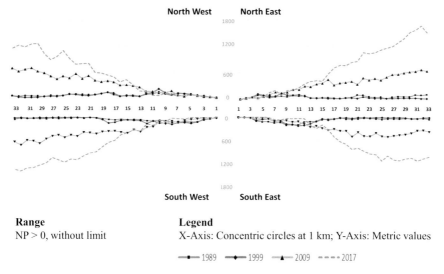

FIGURE 4.5
Number of patch metrics.

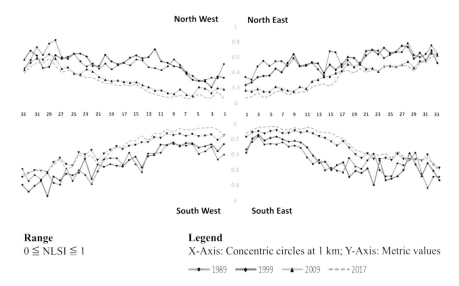

FIGURE 4.6
Normalised landscape shape index.

3. Clumpiness deals with aggregation and disaggregation for adjacent urban patches. Referring to Figure 4.7, in 1989 the values closer to 0 (circles 27–29 NW, 29–33 SW, 29–31 SE) indicate less compact growth or maximum disaggregation. In 2017 the values approaching +1 in all directions, especially in core city areas (circles 1–17), indicate the

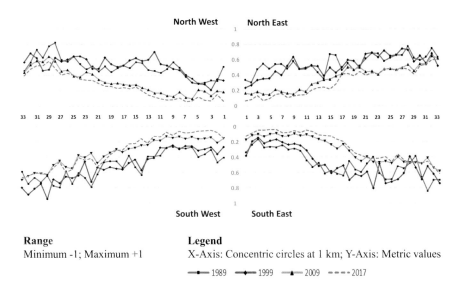

FIGURE 4.7
Clumpiness metric.

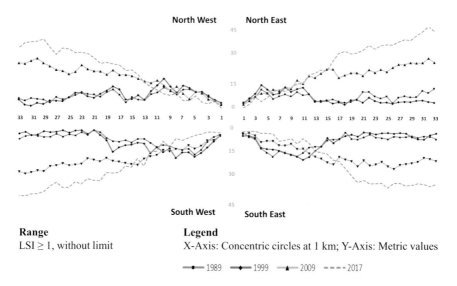

FIGURE 4.8
Landscape shape index.

growth is very complex and all the patches are maximally aggregated to form a large urban monotype patch.

4. The landscape shape index (LSI) is more similar to NLSI, but it provides a standardised measure of total edge or edge density that
adjusts for the size of the landscape. As depicted in Figure 4.8, LSI values are increasing in all directions for Hyderabad during 2009–2017.
The increase in LSI value shows that the landscape shape is becoming
more irregular as the length of edge within the landscape increases.

5. Patch density (PD) is also an indicator of landscape fragmentation. PD
has a linear relationship with NP, as NP increases PD also increases.
PD showed that the values were high in the 1990s, whereas they were
lower in the 2010s. Figure 4.9 represents that the urban patches were
more compact in the core areas and gradually increased, away from
the core area, demonstrating the level of fragmentation in the landscape. Low PD values observed in the core area and the fragmented
buffer zones have higher values during 2017, emphasising the intense
urbanisation at the city centre and sprawl at the outskirts.

6. The aggregation index (AI) is calculated from an adjacency matrix
showing different types of patches appearing side by side or adjacent
to each other in the land use map. The results mainly depict (Figure
4.10) that AI values are less at the outskirts showing maximum disaggregation (years 1989 and 2017 circles 27–33, all four directions).
Values range as high as 80–100 in the year 2017, presenting different
land use types appearing adjacent to the urban category are much

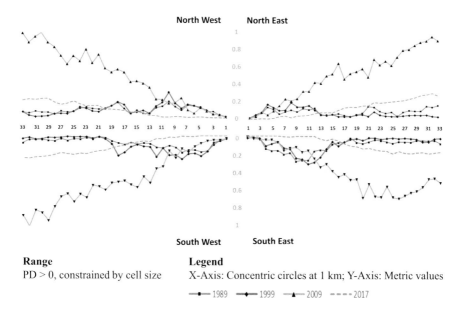

FIGURE 4.9
Patch density metric.

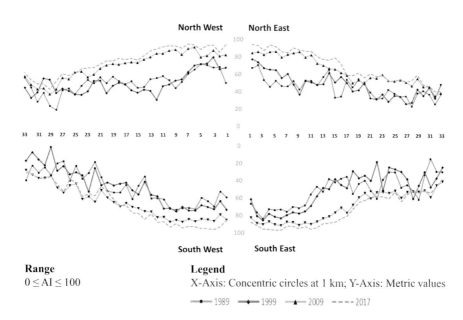

FIGURE 4.10
Aggregation index.

less, i.e., patch type is maximally aggregated into a single and compact patch.

7. Largest patch index (LPI) represents the largest patch and assesses urban land use transitions through different time periods of the study. The largest urban patch was observed in the year 2009 (circles 8–12 NE, 4–11 SW, 5–14 SE) showing the dominance of the lone urban landscape patch type. LPI values decrease from core urban areas to the outskirts, since the largest patch of the urban landscape is increasingly small. Figure 4.11 shows variations in LPI values.

8. The interspersion and juxtaposition index (IJI) shows how well an urban patch is associated or interspersed with other adjacent patch types. Lower values, as observed (Figure 4.12) in 1990s, indicate an urban patch is associated only with one other adjacent patch type. This phenomenon does not hold well at the outskirts since an urban patch is equally adjacent to all other patch types (i.e., maximally interspersed and juxtaposed to other patch types) showing sprawl in these areas.

4.6 Conclusion

This chapter presents an integrated method using classified remote sensing data with gradient analysis, zonal analysis and spatial metrics to understand

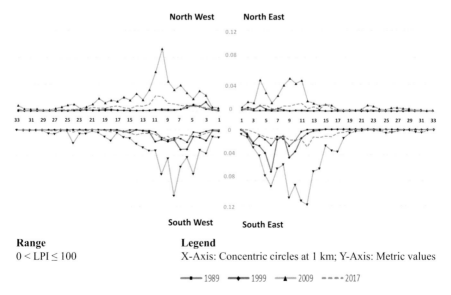

FIGURE 4.11
Largest patch index.

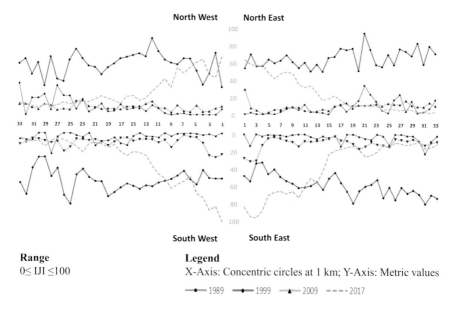

FIGURE 4.12
Interspersion and juxtaposition index.

the spatial temporal patterns of land uses. Further, the study also proves that spatial metrics are excellent tools for detecting spatial patterns of rapid land use transition dynamics, which provide vital insights to environmental conditions of a region through monitoring land changes, identifying gaps between growth pattern and actions taken by local planning authorities. Application of urban spatial metrics, by considering Hyderabad city, India, as an example, have explored the full potential of spatial characterisation of dynamic changes of complex urban systems. Analysed results also reveal a general trend of fragmented urban growth patterns over the years in the periphery of HMDA suggesting strong industry-oriented planning and development activities. The growth of the urban region was rapid with the erosion of vegetation and other land use types. Studies have also shown the importance of spatial metrics in understanding land use dynamics required for urban modelling.

References

Anderson, J. R. 1971. Land use classification schemes used in selected recent geographic applications of remote sensing. *Photogrammetric Engineering*, 37(4), 379–387.

Angel, S., Parent, J., & Civco, D. 2007. Urban sprawl metrics: An analysis of global urban expansion using GIS. In *Proceedings of ASPRS 2007 Annual Conference*, Tampa, Florida May (Vol. 7, No. 11). Citeseer.

Bharath, H. A., & Ramachandra, T. V. 2013. Measuring urban sprawl in Tier II cities of Karnataka, India. In *2013 IEEE Global Humanitarian Technology Conference: South Asia Satellite (GHTC-SAS)* (pp. 321–329). IEEE.

Bhatta, B., Saraswati, S., & Bandyopadhyay, D. 2010. Quantifying the degree-of-freedom, degree-of-sprawl, and degree-of-goodness of urban growth from remote sensing data. *Applied Geography*, 30(1), 96–111.

Congalton, R. G., & Green, K. 2008. *Assessing the accuracy of remotely sensed data: Principles and practices.* CRC press.

Forman, R. T. T., & Godron, M. 1986. *Landscape ecology.* Wiley Publications.

Furberg, D., & Ban, Y. 2012. Satellite monitoring of urban sprawl and assessment of its potential environmental impact in the Greater Toronto Area between 1985 and 2005. *Environmental Management*, 50, 1068–1088.

Galster, G., Hanson, R., Ratcliffe, M. R., Wolman, H., Coleman, S., et al. 2001. Wrestling sprawl to the ground: Defining and measuring an elusive concept. *Housing Policy Debate*, 12(4), 681–717.

Gustafson, E. J. 1998. Quantifying landscape spatial pattern: What Is the State of the Art? *Ecosystems*, 1, 143–156.

Hepinstall-Cymerman, J., Stephan, C., & Hutyra, L. R. 2013. Urban growth patterns and growth management boundaries in the Central Puget Sound, Washington, 1986–2007. *Urban Ecosystems*, 16, 109–129.

Herold, M., Goldstein, N. C., & Clarke, K. C. 2003. The spatiotemporal form of urban growth: Measurement, analysis and modeling. *Remote Sensing of Environment*, 86(3), 286–302.

Herold, M. Couclelis, H., & Clarke, K. C. 2005. The role of spatial metrics in the analysis and modeling of urban change. *Computers, Environment, and Urban Systems*, 29, 339–369.

Hurd, J. D., Wilson, E. H., Lammey, S. G., & Civeo, D. L. 2001. Characterization of forest fragmentation and urban sprawl using time sequential Landsat Imagery. In *Proceedings of ASPRS Annual Convention*, St. Louis, MO.

Jain, S., Kohli, D., Rao, R. M., & Bijker, W. 2011. Spatial metrics to analyse the impact of regional factors on pattern of urbanisation in Gurgaon, India. *Journal of the Indian Society of Remote Sensing*, 39(2), 203–212.

Jim, C. Y., & Chen, W. Y. 2009. Diversity and distribution of landscape trees in the compact Asian city of Taipei. *Applied Geography*, 29(4), 577–587.

Kong, F., Yin, H., Nakagoshi, N., & James, P. 2012. Simulating urban growth processes incorporating a potential model with spatial metrics. *Ecological Indicators*, 20, 82–91.

Leitao, A. B., & Ahern, J. 2002. Applying landscape ecological concepts and metrics in sustainable landscape planning. *Landscape and Urban Planning*, 59(2), 65–93.

Li, X., Jongman, R. H., Hu, Y., Bu, R., Harms, B., et al. 2005. Relationship between landscape structure metrics and wetland nutrient retention function: A case study of Liaohe Delta, China. *Ecological Indicators*, 5(4), 339–349.

McGarigal, K., & Marks, B. 1995. *Fragstats – Spatial pattern analysis program for quantifying landscape structure.* Forest Science Department, Oregon State University.

O'Neill, R. V., Krummel, J. R., Gardner, R. E. A., Sugihara, G., Jackson, B., et al. 1988. Indices of landscape pattern. *Landscape Ecology*, 1(3), 153–162.

Ramachandra, T. V., Bharath, H. A., & Durgappa D. S. 2012. Insights to urban dynamics through landscape spatial pattern analysis. *International Journal of Applied Earth Observation and Geoinformation*, *18*, 329–343.

Ramachandra, T. V., Bharath, H. A., & Sowmyashree, M. V. 2014. Urban footprint of Mumbai - The commercial capital of India. *Journal of Urban and Regional Analysis*, *6*(1), 71–94.

Schneider, A., Seto, K. C., & Webster, D. R. 2005. Urban gowth in Chengdu Western China: Application of remote sensing to assess planning and policy outcomes. *Environment and Planning B*, *32*, 323–345.

Sudhira, H. S., Ramachandra, T. V., & Jagadish, K. S. 2004. Urban sprawl: Metrics, dynamics and modelling using GIS. *International Journal of Applied Earth Observation and Geoinformation*, *5*(1), 29–39.

Sun, C., Wu, Z. F., Lv, Z. Q., Yao, N., & Wei, J. B. 2013. Quantifying different types of urban growth and the change dynamic in Guangzhou using multi-temporal remote sensing data. *International Journal of Applied Earth Observation and Geoinformation*, *21*, 409–417.

Taubenbock, H., Wegmann, M., Roth, A., Mehl, H., & Dech, S. 2009. Urbanization in India-Spatiotemporal analysis using remote sensing data. *Computers, Environment and Urban Systems*, *33*(3), 179–188.

Thapa, R., & Murayama, Y. 2009. Examining spatiotemporal urbanization patterns in Kathmandu Valley, Nepal: Remote sensing and spatial metrics approaches. *Remote Sensing*, *1*(3), 534–556.

Thomas, R. W. 1981. *Information statistics in geography*. Norwich, UK: Geo Abstracts, University of East Anglia.

Torrens, P., & Alberti, M. 2000. *Measuring sprawl*. CASA working paper series 27. Accessed on November 6, 2018. Retrieved from http://www.casa.ucl.ac.uk

Tsai, Y. H. 2005. Quantifying urban form: Compactness versus' sprawl'. *Urban Studies*, *42*(1), 141–161.

Turner, M. G., O'Neill, R. V., Gardner, R. H., & Milne, B. T. 1989. Effects of changing spatial scale on the analysis of landscape pattern. *Landscape Ecology*, *3*(3–4), 153–162.

UNDESA. 2011. *Population distribution, urbanization, internal migration and development: An international perspective*. United Nations publication. Accessed on December 28, 2018. Retrieved from www.unpopulation.org

Wu, J. 2006. Environmental amenities, urban sprawl, and community characteristics. *Journal of Environmental Economics and Management*, *52*(2), 527–547.

5

Land Use Modelling

Future Research, Directions and Planning

CONTENTS

5.1 Introduction 81
5.2 Approaches to Model LULCC and Evolution of Urban Models 85
5.3 Cellular Automata Models 86
5.4 Application of CA with Markov Process... 88
5.5 Literature Survey on CA Models.. 90
  5.5.1 CA-Fuzzy Model 91
  5.5.2 CA–Analytical Hierarchical Process (AHP) and Multi-
       Criteria Evaluation (MCE) Model 91
  5.5.3 CA-Markov Model 92
  5.5.4 CA-Based SLEUTH Model 92
       5.5.4.1 Improvements and Modifications of SLEUTH 94
       5.5.4.2 SLEUTH Input Data Set Preparation 95
       5.5.4.3 Model Test, Calibration and Prediction 96
5.6 Visualisation of Urban Growth (2025) 96
5.7 Predicting Future Urban Expansion: The Case of Ahmedabad 98
5.8 Improving SLEUTH Model Calibration through GA............................ 99
5.9 Limitations/Drawbacks of CA models.. 101
5.10 Concluding Remarks 102
References.. 102

5.1 Introduction

This chapter introduces the tenets of landscape modelling and application with a case study of cellular automata models. Further, the last section of this chapter deliberates on the prediction with general model structure details. It has been over two decades since the inception and implementation of a land use and land cover change (LULCC) project collaborated jointly by the International Geosphere-Biosphere Program (IGBP) and the International Human Dimensions Program on Global Environmental Change (IHDP). The LULCC project aimed to address LULC dynamics through the lens of developing common models, considering local level agents or drivers of change. Humans have been altering landscape structures to meet their needs. However, unplanned land use changes have affected the ecological functions of a landscape, evident from barren hilltops, conversion of perennial streams to seasonal streams and profuse spread of invasive species, which have been threatening the sustenance of natural resources affecting the food and water security of a region. This necessitates the analyses of LULCC to evolve strategies for prudent planning and sustainable management of natural resources and mitigation of changes in the climate. Mitigation of impacts of LULCC can be addressed through simulation and geovisualisation of likely changes through robust land use models. Essential features of robust land use models are (1) they are built on proper scientific background along with supporting data sets; (2) they incorporate technical innovations to provide realistic results; and (3) the results are repeatable, replicable and robust, and (4) transparent, flexible and adaptable to local and regional policies. Challenges faced due to model land cover changes are manifold since global level (top-down approach) modifications considering socio-economic and biophysical factors could severely affect local levels. This prompts for an effective scenario framework at the respective scale to foresee impending developments occurring in existing complex ecosystems. It is extremely difficult to predict long-term land cover change with minimum uncertainty because of the dynamics governing LULCC. In this case, scenario-based prediction helps us understand various possible facets of land use growth and change in the next couple of decades. Figure 5.1 depicts three different scenarios indicating urban land use changes over the years. These kinds of models help in exploring maximum possible conditions on the ground by considering a wide range of factors effecting LULCC. Use of scenarios helps in designing policies in favour of human well-being keeping in mind geomorphology (environment), connectivity (mobility), facilities, government (development plans), demographic, socio-economic, sustainability (waterbodies, green spaces, etc.) and other factors with their effects on LULCC (Wahyudi & Liu, 2013). Figure 5.2 provides detailed illustration of factors driving urban growth responsible for rapid land use changes. Scenarios may be region or country specific, since there are enormous factors that effect changes in land use in the future.

For instance, let us consider urban land use dynamics in a specific region and its prediction until the year 2090 based on historical (previous and present) data sets. The urban land use is 48% as of 2018, depicted in Figure 5.2. Prediction of likely land uses is done considering four different scenarios: Scenario A gives the business as the usual trend; the model works simply based on past inputs and present conditions, and there would be no restrictions in growth. Scenario B is assumed to have linear growth with limited restrictions. Scenario C would see decline in urban areas due to strict measures and environmental guidelines, specifically aiming at protection of natural green cover and waterbodies. Scenario D projects decline after the year 2060 and this case is very unlikely to happen.

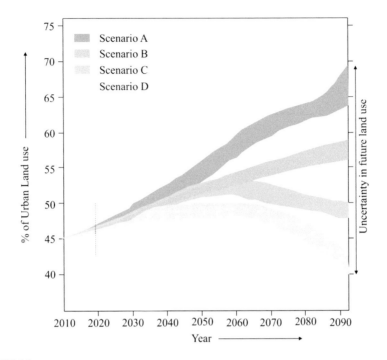

FIGURE 5.1
A scenario-based model prediction for change in land use. (Figure adapted from Sohl &
Sleeter, 2012.)

Visualisation of future trends entails development of data sets for models such as historical data, influencing factors, drivers and constraints
affecting land use changes. A model is an abstract form or simplification
of the real world. Modelling uses artificial representation and interaction
between human activities and land use systems to successfully mimic future
development.

The primary aim of models is to understand, define, quantify, visualise
or simulate a future scenario by referring to previous and present data sets.
Since it is difficult to implement any kind of model on a real-scale experiment, researchers provide near-realistic computation models with hypoth-
eses testing of the process of LULCC. Models help to explore situations
through what-if analysis, and the visualisation obtained helps planners to
make decisions in the best interest of the environment. Apart from land
use/cover modelling, models are used in various fields such as hydrological
modelling, climate change modelling, environment modelling and ecosystem modelling. One such common land use change estimation model is the
urban growth model. It has been proposed and used for over four decades
for testing theories related to spatial location and interaction between land
use categories as well as associated activities. Urban models also consider
and represent a framework based on population, land use, transportation

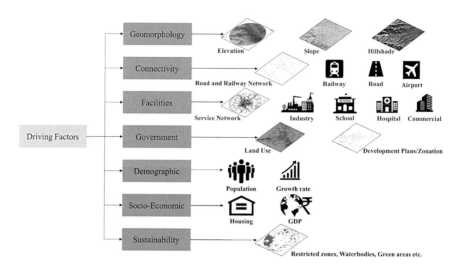

FIGURE 5.2
Urban growth driving factors considered by various researchers for developing models. (Adapted from Wahyudi & Liu, 2013.)

and other socio-economic factors (Batty, 2009). These models mainly concentrate on allocation of different land use activities within the specified boundaries. The general structure of LULCC models is shown in Figure 5.3. The method includes three key aspects: (1) data acquisition and pre-processing; (2) land use and land cover analysis using suitable algorithms/classifiers; and (3) model input, sensitivity analyses and validation. Typically, visualisation through maps is an end product of modelling.

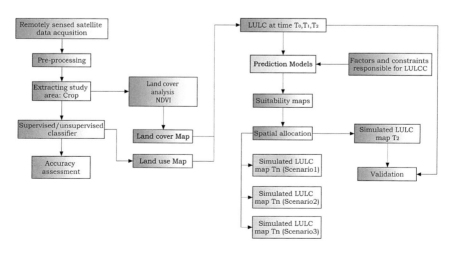

FIGURE 5.3
General method used to predict future LULCC.

5.2 Approaches to Model LULCC and Evolution of Urban Models

LULCC models can be broadly divided into eight categories based on their characteristics (Parker et al., 2003). They are as follows:

1. Mathematical equation-based models – Linear programming model. Spatial aspect is difficult to integrate.
2. System models – Designed based on stock and flow data as a set of differential equations, depending on functional relationships.
3. Statistical techniques – Includes regression techniques. Comparatively easier to integrate spatial data by treating raster images as a matrix.
4. Expert models – Combines expert judgement along with probability techniques. Provides occurrence of likely land use changes.
5. Evolutionary models – Extensively use evolutionary algorithms and artificial intelligence.
6. Cellular models – Include cellular automata and Markov chains. Operate based on cell structure and rules. Extremely powerful in both spatial and temporal aspects.
7. Hybrid models – Combine any of the aforementioned models, taking advantage of each model and expected to deliver better results.
8. Agent-based models – Agents have control over their actions to achieve dedicated goals. Capable of providing a realistic scenario.

Further, LULCC model classification based on large diversity are of six types: spatial versus non-spatial, dynamic versus static, descriptive versus prescriptive, deductive versus inductive, agent-based versus pixel-based representations and global versus regional models (Verburg et al., 2006). When it comes to adaptation of the best suitable LULCC model, there is no universal approach. Choice of model therefore depends on research questions to be answered, data availability, factors considered, region, extent of spatial and temporal resolutions, type of land use changes to be addressed, etc. For example, cellular automata (CA)-oriented models are best suited for urban growth dynamics, but they fail to determine agriculture land transitions due to a large number of sub-classes within itself. Urban models have gained popularity due to their effectiveness in addressing dynamic changes. Urban modelling is the process of identifying a theory that could be translated into a mathematical model as well as developing specific computer-aided programs to feed the model with data so as to calibrate, validate, verify and predict future urban trends (Batty, 2009). The scientific community over the last four decades has been contributing immensely to urban growth models having a common goal to study land use dynamics and simulate urban growth using geospatial techniques (EPA, 2000; Sante et al., 2010). Theoretical assumptions, method followed, spatial and temporal aspects, and geographical extents might vary from each model type, but the final outcome of these models are to understand the complex interrelationships between the natural ecosystem and urban environment by observing irreversible heterogeneous patterns of change (Yeh & Li, 1998; Li & Liu, 2006). Wu and Silva (2010) highlight the significance of artificial intelligence (AI) and its deep theoretical understanding of the urbanisation process and relative pressure on land use pattern change based on the review of urban land dynamics and related models. Figure 5.4 shows broad categories of AI systems used in urban models. This chapter focuses on the cellular automata and cellular automata–based SLEUTH models.

5.3 Cellular Automata Models

The origin of CA-based models dates back to the 1940s, with an investigation of the CA framework, complex behaviour of systems, and its effect and functional aspects (Ulam, 1958; Torrens, 2000). Based on the theoretical concepts

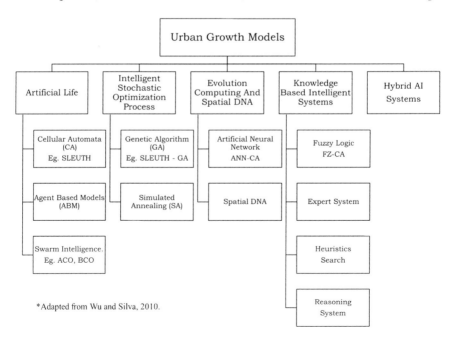

FIGURE 5.4
Different types of urban growth models.

laid by Von Neumann and Ulam, Conway developed the simulation game "life" (Gardner, 1970) and Wolfram (1986) developed algorithms and studied rules of one-dimensional CA having diverse applications. CA has been considered and extensively utilised by the scientific community due to its simplicity, flexibility and intuitiveness in the majority of the fields such as medicine, biological modelling, forest fire modelling, flood modelling, and urban and landscape dynamics (Ermentrout & Edelstein, 1993; Li & Gong, 2016).

The progression of CA coupled with geographic modelling started with an understanding of the states of cells over a specified time and set of rules (Tobler, 1979). A theoretical framework was designed to simulate urban growth, followed by studies conducted by several well-known researchers who did extensive work on approaches to CA models for simulation (Couclelis, 1985; White & Engelen, 1994). CA models can be defined as a class of two-dimensional lattice of cells, where each cell representing a land use category tends to change its state over time depending on the local neighbourhood of every cell. A brief categorisation of CA-based urban growth models (UGMs) can be seen in Figure 5.4. Standard cellular automata apply a bottom-up approach. The approach argues that local rules can create complex global patterns by running the models in iterations (Batty & Xie, 1994). Itami (1994) states that CA is seen not only as a framework for dynamic spatial modelling but as a paradigm for thinking about complex spatial–temporal phenomena and an experimental laboratory for testing ideas.

Each cell (automaton) S is defined as

$$S^{t+1} = f\left(S^t, N\right) \tag{5.1}$$

where S is a set of all possible states of the cellular automata, N is a neighbourhood of all cells providing input values for the function f, and f is a transition function that defines the change of the state from t to $t + 1$. CA are discrete two-dimensional dynamic systems in which local interactions among components generate global changes in space and time (Wolfram, 2002) and are composed of four components (Torrens, 2000), namely:

1. Cell space represented by an array of cells; these cells may be in any geometric shape.
2. A number of finite states that qualifies the state of each cell.
3. Neighbourhood.
4. Transition function, which define the next state of the cell in the next time period, based on the given state of the cell itself and its neighbourhood cells.

Basic entities of CA:

1. Cell – These are the basic elements of an automata (behaving like a memory unit and stores binary states such as 0 and 1).

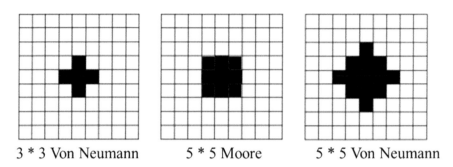

3 * 3 Von Neumann   5 * 5 Moore   5 * 5 Von Neumann

FIGURE 5.5
Different types of neighbourhoods. (Adapted from Jahan & Khosrojerdi, 2016.)

2. Lattice – It is formed when cells are arranged in a spatial web.

3. Rules – Cells in a lattice represent a static state. To introduce a dynamic system, rules are necessary and help in defining a state of the cell for the next time in dependence of the neighbourhood cells.

4. Neighbourhood – Consists of adjacent cells surrounding a central cell on a two-dimensional square lattice. There are different kinds of neighbourhoods that differ mainly in size and shape (explained in Figure 5.5):

   a. Von Neumann – The von Neumann neighbourhood comprises the four cells orthogonally surrounding a central cell on a two-dimensional square lattice.

   b. Moore – It includes the von Neumann neighbourhood as well as the four remaining cells surrounding the cell whose state is to be calculated (Kier et al., 2005)

In the nearest-neighbour model, there are actually only $2^3 = 8$ different configurations of neighbours and the site itself leading to the outcome. These are

111   110   101   **100**   **011**   010   001   000

For each of the eight states, there are two possible outcomes, i.e., either 0 or 1. If we consider the following example as a rule, the state of cells in time step $t + 1$ is given according to "Rule 11111010"

t:	111	110	101	100	011	010	001	000
$t + 1$:	1	1	1	1	1	0	1	0

5.4 Application of CA with Markov Process

A Markov process is a stochastic process (random process) in which the probability distribution of the current state (t) is conditionally independent of the past states $(t - 1)$; to predict future states $(t + 1)$ depends on the state at time $t(i_t)$. This as a characteristic is called the Markov property. A discrete-time stochastic process is a Markov chain if, for $t = 0, 1, 2, \ldots$ and all states are defined as shown in Equation 5.2:

$$P\left(X_{t+1} = i_{t+1} \middle| X_t = i_t, X_{t-1} = i_{t-1}, \ldots, X_1 = i_1, X_0 = i_0\right) = P\left(X_{t+1} = i_{t+1} \middle| X_t = i_t\right) \quad (5.2)$$

Assuming that for all states i and j as well as all t, $P(X_{t+1} = j | X_t = i)$ is independent of t. According to the assumption the equation would be

$$P\left(X_{t+1} = j \middle| X_t = i\right) = P_{ij} \quad (5.3)$$

where P_{ij} is the probability given to the system, in state i at time t. It will be in a state j at time $t + 1$. Figure 5.6a shows a transition probability matrix, where the probabilities are displayed as $s \times s$. The transition probability matrix records the probability that each land cover category will change to every other category. The transition areas matrix records the number of pixels that are expected to change from each land cover type to each other land cover type over the specified number of time units. The Markov chain controls the temporal dynamics, whereas CA controls the spatial dynamics aspects.

CA-Markov uses cellular automata procedures in combination with Markov chain analysis and multi-criteria evaluation (MCE). This process integrates transition suitability obtained from the Markov chain, number of iterations to be performed and filter to be used to predict future land use. The number of iterations can be specified by the user. A filter is a spatially explicit contiguity-weighting factor and works on the principle that a pixel close to a specific land use is more likely to change to that category than a distant one (Figure 5.6b). These filters essentially are kernel, implying a linear reduction of weight within neighbouring cells with distance. Various studies have incorporated an integrated approach of the aforementioned models while

$$P = \begin{bmatrix} P_{11} & P_{12} & \cdots & P_{1s} \\ P_{21} & P_{22} & \cdots & P_{2s} \\ \vdots & \vdots & & \vdots \\ P_{s1} & P_{s2} & \cdots & P_{ss} \end{bmatrix}$$

(a)

0	0	1	0	0
0	1	1	1	0
1	1	1	1	1
0	1	1	1	0
0	0	1	0	0

(b)

FIGURE 5.6
(a) Transition probability matrix. (b) A 5×5 contiguity filter.

considering geographic, topographic and socio-economic factors to predict future land use changes in different parts of the world (Tang et al., 2007; Lu et al., 2009; Chen et al., 2010; Mahiny & Clarke, 2012).

5.5 Literature Survey on CA Models

Couclelis (1997) demonstrated the integration of GIS and CA models with simulations based on actual data of a real-time scenario. He tries to explain the solution for theoretical problems with respect to proximal space and practical applications, which can be handled successfully using geo-algebra. Gradually, research work started focusing on practical applications from theoretical approaches. Some of the first applications of urban CA to the simulation of real-world cases were carried out by Batty and Xie in 1994. The authors highlight the applicability of CA models considering the city of Savannah, Georgia.

Ward et al. (2000) conducted a descriptive study with 50 m cell size for quantifying urban and non-urban areas based on Moore and von Neumann neighbourhoods. Population growth was taken as a constraint while predicting changes. The authors set up a transition rules where the state of the cell is said to be urban if the cell is not affected by constraints, or if at least one of the neighbouring cells belongs to the transportation network or if the value of a random variable is greater than the specified growth rate and if there is no directional bias in the location of the cell in the neighbourhood.

Barredo et al. (2003) proposed a bottom-up approach integrating CA for modelling future urban land use and also analysed the role of urban land use allocation factors. Simulation for Dublin was carried out for the period 1968–1998 to explain the practicality of CA, by comparing the simulated map with actual land use in the city through (a) visual comparison, (b) fractal dimension and (c) a co-incidence matrix.

Further He et al. (2006) examined an urban expansion scenario (UES) model by integrating the cellular automata model and system dynamics (SD) model to understand the continuous urban expansion versus water resource restriction and other environment constraints in Beijing Municipality. Land use maps were produced using Landsat TM and ETM+ data for the years 1991, 1997, 2000 and 2004. A simulated image of 2004 was validated and further used for predicting land use for the years 2007, 2010, 2015 and 2020 considering four cases, i.e., without restriction of urban planning policy, with restriction of high-yield farmland protection area, with restriction of green belt, and with restriction of both the green belt as well as high-yield farmland protection area. CA was successful in modelling urban expansion patterns to keep the balance of the urban land demand and supply.

Bharath et al. (2013) used a land change modeller (LCM) with a Markov–cellular automata model along with various agents of development for one of the most rapidly urbanising cities in India – Bangalore – to predict land use changes for the year 2020 using classified images during the period 2006–2012. Modelling was carried out by considering a waterbody as a constraint. Comparison of the classified map (for 2012) with the predicted map with a kappa value of 0.91 and overall accuracy of 93.64 and hence the technique was repeated for predicting the urban growth. This model's results showed increase in the urban category and severe decrease in the vegetation category, which was an indicator of urban land use reaching its maximum threshold.

5.5.1 CA-Fuzzy Model

Zadeh (1965) defines a fuzzy set as a class of objects with a continuum of grades of membership. Such a set is characterised by a membership function which assigns to each object a grade of membership ranging between zero and one, compared to traditional binary logic where variables may take either true or false values. Fuzzy logic is a form of many-valued logic that deals with approximate rather than fixed and exact reasoning. Urban modelling incorporating fuzzy logic governs the land use conversion process where state of the cell or pixel changes are simulated by fuzzy sets that describe preconditions of an action (Wu, 1998).

Wu (1998) developed a framework based on a computer-based approach for simulating land encroachment using a classified map derived from Landsat TM-5 with fuzzy-logic-controlled (FLC) cellular automata. FLC also allows modifications of transition rules making it easier to understand. The effort involved 40 iterations, and for each simulation, four different simulations were observed: baseline scenario, highway promoted development, without any constraint on woodland and without any constraint on agriculture area. Wu concluded that FLC integrated with CA and GIS was successful in simulating urban edge expansion, which further gave insights to improve future policy measures.

5.5.2 CA–Analytical Hierarchical Process (AHP) and Multi-Criteria Evaluation (MCE) Model

The analytical hierarchical process (AHP) was initially developed by Saaty (1980) and uses a pairwise comparison matrix created by setting out one row and one column for each factor (Satty & Vargas, 2001). AHP allows combinations of group judgements and satisfies the reciprocal property for the group by comparing two factors (Thapa & Murayama, 2009). The CA-AHP model has a stronger capacity for interpretation with the consideration of dynamic multi-criteria evaluation (MCE). Adjustment of factor weights through AHP would provide distinctive scenarios.

   Bharath et al. (2014) conducted a detailed study on urban landscape dynamics, sprawl and modelling by combining CA, Markov, fuzzy-AHP and MCE. The study demonstrates MCE accommodating factors and their weights derived from AHP for different agents such as road networks, industries, educational institutions and bus stops, and constraints such as Boolean were used to generate suitability maps for different land use classes. Fuzzyfication was adopted to bring all factors on a same scale of 0–255, where 0 indicated least or no changes and 255 indicated highest probability of change of land use from one category to another. Based on factors and constraints, two scenarios were observed, i.e., prediction of land use map for the years 2016 and 2020 with the constraint of a city development plan (CDP) and without a CDP. The predicted land use without a CDP showed intensified growth well within city limits severely impacting the local ecology and environment.

5.5.3 CA-Markov Model

Markov chains are the simplest mathematical models for random phenomena evolving in time. CA-Markov is a combined cellular automata/Markov chain/ multi-criteria land cover prediction procedure that adds an element of spatial contiguity as well as knowledge of the likely spatial distribution or in other words the probability of transitions to Markov chain analysis (Eastman, 2009).

   Bharath et al. (2013) explores the capabilities of the Markov chain and cellular automata model with the help of temporal Landsat data for the years 2008, 2010 and 2012. The Markov module was used to obtain a land use change probability map and land use change probability areas for in the Bangalore region, whereas CA helped to obtain a spatial context and distribution map. Besides incorporating CA transition rules, researchers also claim that a few constraints were taken to predict land use changes for the year 2020 along with spatial metrics and analysis, which gave more insights in identifying urban growth patterns. Several investigators have integrated a Markov chain and CA urban growth model considering various socio-economic factors to predict and analyse future expansion and effects of urbanisation in India.

   Silva and Clarke (2002) insist on the importance of validating the modelling process. They outline validating the urban expansion scenario before using the model to predict future land use patterns since the simulated urban growth pattern not only depends on urban land demand but it is also greatly influenced by the factors and weights associated with driving and resisting forces or constraints at local levels. Validation is possible only when urban land use is known for a 2-year time period in history with a significant gap in between (Verburg et al., 1999).

5.5.4 CA-Based SLEUTH Model

The history of agent-based models (ABMs) dates back to the 1950s when Stanislaw, Ulam and von Neumann created simple two dimensional discrete cellular automata (Wolfram, 2002). Potential usage of ABM in research areas like LULCC simulation, design, development and implementation has been explored (Parker et al., 2003; Sun et al., 2014). It is necessary to acknowledge the importance of both spatial and temporal dynamics to predict land use changes. While CA can capture and control the spatial dynamics aspects, Markov chain integrated controls the temporal dynamics of a region. In this aspect, the SLEUTH model takes into account both spatio-temporal aspects of urban growth trajectories for simulation and prediction. The model was developed by Clarke, completely based on cellular automation along with terrain mapping and land cover deltatron. The acronym of SLEUTH stands for slope, land use, excluded, urban, transportation and hillshade, the layers used as input for the model, and consists of C-language, using UNIX or UNIX-based operating systems (Rafiee et al., 2009; Chandan et al., 2019). SLEUTH has the ability to predict urban/non-urban land use dynamics based on two submodels: urban growth model (UGM) and deltatron land model (DLM) (Dietzel & Clarke, 2004). Specifications to run the model are demonstrated by Silva and Clarke (2005) as a grid space of homogeneous cells, with a Moore neighbourhood of eight cells, two different cells (urban/non-urban states), and, most important, the five factors controlling the behaviour of growth, namely diffusion, breed, spread, slope resistance and road gravity. Apart from these factors, the model also considers growth rules categorised into four broad types: spontaneous growth, new spreading centre growth, edge growth and road-influenced growth. Figure 5.7 depicts the interrelationship between factors and growth rules. The factors can be calibrated according to the input layers provided and region under study.

The first research on the SLEUTH model started in the year 1997. Clarke et al. (1997) were successful in designing initial stages of SLEUTH to build

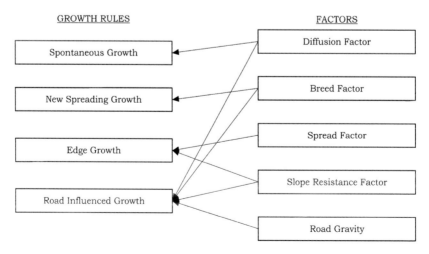

FIGURE 5.7
Interrelationship between growth rules and factors.

and calibrate the model and visualise historical urban growth as a part of a research project under USGS in collaboration with Hunter College (New York City). They demonstrated the use of self-modifying CA for the San Francisco Bay Area from 1900 to 1990, and simulated growth under three different prediction modes: uncontrolled rapid growth, sustained slow growth and limited/desirable growth. The authors have tried to address problems faced during the calibration phase by allowing the growth rate of complex curves under self-modifying CA rather than limiting it to linear or exponential curves followed by the traditional CA method.

Research was extended to the Washington/Baltimore corridor by Clarke and Gaydos (1998). They employed a procedure with initial calibration of 7560 combinations executed in 252 CPU hours followed by 3000 combinations in 100 CPU hours. The final run of 100 Monte Carlo (MC) iterations was performed to predict future transition until the year 2100. A conclusion was drawn that repeated a new spreading centre formed during numerous model trials indicating that the city is ready to accept a high urban growth rate in the near future.

The SLEUTH model has been extensively used because of its excellent compatibility, availability of open source files with the demo data sets to execute the program (Project Gigalopolis) and better accuracy testing methods. Wu et al. (2009) explored the potential application of the model to the Shenyang region in China. They collected urban and road data for the years 1988, 1992, 1997, 2000 and 2004 using various geospatial sources along with exclusion, slope, hillshade and mask layers as input to the model. They used optimal coefficients to achieve desired accuracy levels by initialising urban extent of the year 1988 data and finally predicting urban growth for the year 2004 using a hundred MC iterations. The model output urban extents for the years 1992, 1997, 2000 and 2004 were compared with satellite-based classified images to arrive at statistical model validation and further prediction.

5.5.4.1 Improvements and Modifications of SLEUTH

Open source SLEUTH has witnessed numerous applications over different parts of the world. Most researchers have successfully attempted to reduce SLEUTH computation time and therefore increase its efficiency. The original version includes SLEUTH UGM and DLM. To reduce computation time, Dietzel and Clarke (2007) worked on the development of an optimisation of the SLEUTH metric called optimal SLEUTH metric (OSM). OSM narrows the parameter range and returns a single goodness-of-fit metric. OSM is the product of various individual metrics like compare, population, edges, clusters, slope, X-mean and Y-mean providing reliable calibration results. The source code for OSM is also available in the Project Gigalopolis website and one can readily implement the OSM technique. Sakieh and Salmanmahiny (2016) explored OSM and implemented it to the Gorgan area of Iran. The authors mentioned the term "cancer-treating" to understand and predict the urban spread in three different scenarios: forest protection, rangeland protection and historical growth (without any restrictions). They have also adopted a comparative assessment to measure association between landscape metrics and land suitability values using spearman correlation to address cancer-treating urban growth pattern. Guan and Clarke (2010) developed a parallel version of SLEUTH called pSLEUTH. It uses an open source parallel raster processing programming library (pRPL) to improve model efficiency, computational performance and, most important, to reduce time taken during the calibration process. Results obtained from pSLEUTH produced different "best-fit" parameter combinations. Authors have applied pSLEUTH to predict urban growth in the continental United States with an image size of 4948 × 3108 pixels. They concluded pSLEUTH utilises features of pRPL and it greatly reduces computing time during calibration. However, a limitation of this method is it does not ensure upgradation in the simulation process. Further, Jantz et al. (2004) highlighted several limitations of SLEUTH and created a new version called SLEUTH-3r. Limitations were positively addressed by modification of source code and by dividing the entire study region into various subdivisions and then simulating each of them independently. Successful application of SLEUTH-3r was tested and reported to be computationally more efficient, and memory usage was reduced at least by 65% compared to the traditional method of SLEUTH. The last and the most recent advancement is the SLEUTH-genetic algorithm (GA). Further details of GA are discussed in the next section. Output of GA includes one nearly optimal solution over a set of iterations. In the case of SLEUTH, GA returns a single best-fit statistic for the entire dynamic range consisting of several Monte Carlo iterations. A study conducted by Clarke-Lauer and Clarke (2011) showed 70% of the chromosomes performed better than traditional SLEUTH, with only one-fifth of the computation time and an improved goodness-of-fit measure. SLEUTH has therefore been used by scientists with a wide spectrum of applications throughout the world, for instance in the United States, Portugal, China, Egypt, Mexico, India and European countries.

5.5.4.2 SLEUTH Input Data Set Preparation

The six input layers for SLEUTH model are acquired from various data sources (Table 5.1). These layers are collectively taken into a single directory and ensured of identical rows and columns, same projection, same map extent and standard resolution at 30 m as specified by researchers. Layers have to be compiled, reclassed, cropped, resampled and exported as greyscale 8-bit GIFs in three different resolutions: coarse (120 m), fine (60 m) and final (30 m) for the purpose of three modes of calibration. Obtained layers should be renamed according to specified format in the Gigalopolis website. Each SLEUTH run depends on the scenario file where most of the variables and settings are written in accordance with derived layers.

TABLE 5.1

SLEUTH Input Data and Source

Layer Name	Source
Slope and hillshade	Processed from ASTER DEM (raster)
Land use and urban extent	Classified from Landsat series images (raster)
Transportation (roads)	Street data (OSM, Bhuvan and Google maps) updated with classified images (originally vector, rasterised)
Excluded map	CDP and other plans (originally vector, rasterised)

5.5.4.3 Model Test, Calibration and Prediction

Details of the SLEUTH general structure and workflow are depicted in Figure 5.8. For the model test phase, full-resolution images (30 m) are used. The scenario file holds information about path directory, log file preferences, Monte Carlo iterations, coefficient range and step values, output statistics, etc., and can be modified from source file as per the formatted study region specifications. Coefficient values will be set as default to test the data compatibility with the model. Monte Carlo iterations are fixed to four and the test phase is run.

The calibration phase involves rigorous iterations to narrow the range of best and optimum suitability of coefficient values required to predict urban growth. During coarse calibration, 120 m resolution images are considered for analysis with 5 MC iterations. Coefficient values for diffusion, breed, spread, slope resistance and road gravity can be explored over the entire dynamic range of 0–100 at a time step of 25. This range is further reduced to a time step of 5 during fine calibration. The final calibration mode has full-size resolution or 30 m. The top Lee-Sallee metrics are selected to obtain five unique coefficient values responsible for urban growth. The scenario file will be modified accordingly to predict urban growth. After each run, the model returns a total number of 13 metrics to evaluate the goodness of fit. Various researchers have used the selected combination of these metrics to estimate optimum values for predicting a future urban scenario.

5.6 Visualisation of Urban Growth (2025)

After ground truth validation and comparison of simulated results using the SLEUTH model, the next step is to visualise changes of urban growth for the next decade, say 2025. With already existing industry and affordable housing policies, government is further planning (by 2030) to achieve universal access to reliable and modern energy services, expand infrastructure and upgrade technology for supply of sustainable energy, and build resilient

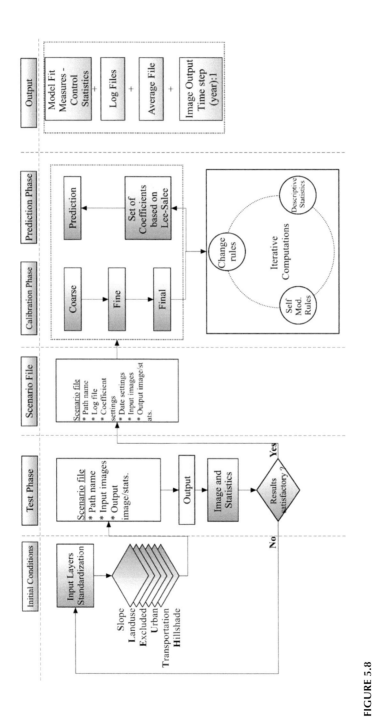

FIGURE 5.8

The SLEUTH structure and workflow. (Adapted from Chaudhuri & Clarke, 2013.)

infrastructure to improve integrated human settlement and rural–urban transport and therefore promote economy. These goals have been asserted by the Government of India emphasising the implementation of Sustainable Development Goals (SDGs) to the United Nations High-Level Political Forum on Sustainable Development (VNNR, 2017). Therefore, visualising urban growth for the year 2025 provides comprehensive strategies and policy revisions in combating unplanned growth to achieve harmony between development and the surrounding natural ecosystem.

5.7 Predicting Future Urban Expansion: The Case of Ahmedabad

Ahmedabad lies on the bank of the river Sabarmati in the northern part of Gujarat at an elevation of 53 m above mean sea level. It is the seventh largest Indian metropolitan city with a population of 6.3 million in 2011 (4.5 million in 2001). The city experiences a hot and dry kind of weather with average temperature ranging between 36°C and 43°C (in summer) and 15°C and 23°C (in winter) with average rainfall of about 1017 mm. Majorly dominated by the textile industry, it is the financial and economic hub of Gujarat, and often is referred to as the "Manchester of India". In 2010 it had a GDP of 59 billion USD as a result of substantial growth of the city's economy. The last two decades saw the city attracting many foreign investments making Gujarat one of the few economically developed Indian states. Ahmedabad Municipal Corporation controls administration of Ahmedabad.

The model input layers are illustrated in Figure 5.9. The results of the calibration phase are discussed. The city showed fewer coefficient values for diffusion (1 and 6) indicating low possibility of overall depressiveness of urban settlements at the outskirts of the Ahmedabad Urban Development Authority (AUDA) boundary. These low values were further justified by breed coefficient (4 and 6) suggesting very minimal chances of newly occurring detached types of urban patches. Inference drawn from the slope coefficient (20 and 1) shows the region has lesser slope resistance to development meaning the area is generally flat. A medium range of spread coefficient (25 and 71) specifies existing urban pockets can influence radiating new settlements attached to them. High values of road gravity coefficient (76 and 67) illustrate the urban growth along major roads and highways. As for the model fit functions (validation) considered, population, edges, clusters and cluster size have shown satisfactorily good results. Regressed values and metrics such as r^2compare and r^2population are used as model fit measures. Cluster and cluster size metrics were close to unity depicting agreement between modelled urban clustering versus known recent urban clustering derived from control years. Modelling results reveal an increase in urban

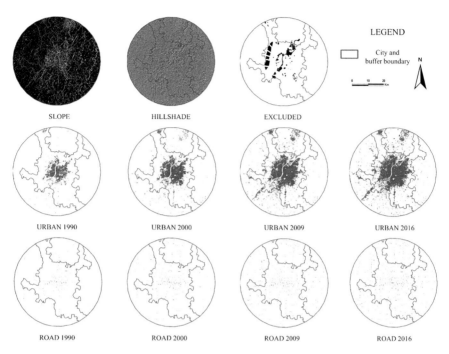

FIGURE 5.9
Input layers used for modelling of Ahmedabad region.

land use from 497.50 km^2 in 2017 to 826.24 km^2 for the year 2025 considering a business as usual scenario. A similar analysis was conducted for other major cities of India, and model prediction results and statistics for the year 2025 are shown in Figure 5.10 and Figure 5.11.

5.8 Improving SLEUTH Model Calibration through GA

SLEUTH-GA is adopted to improve the calibration process and to obtain optimised values in less computation time. Figure 5.12 shows the SLEUTH-GA model calibration approach. Coarse, fine and final calibration procedures employed in the brute force method is replaced by GA calibration, apart from which other procedures remain the same. Strength of GA comes from its ability to explore search space with improved results after iterations. GA applied to SLEUTH follows a four-step process: population initialisation, selection strategies, crossover breeding and finally mutation. Population initialisation works on three strategies in determining seed population: stratified, partial random and random. Selecting chromosomes involves two methods: elitism and tournament selection. The breeding method includes

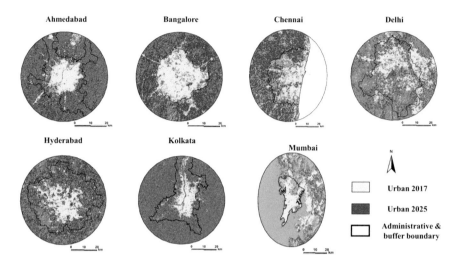

FIGURE 5.10
Simulated urban growth of major Indian cities for the year 2025.

a uniform crossover technique where each gene has a chance of getting into the next-generation child. The mutation stage involves changes in gene number by addition or subtraction of a random number.

The minimum and maximum value for a gene is fixed as 1 and 100, respectively, defining the range of each coefficient in the SLEUTH procedure.

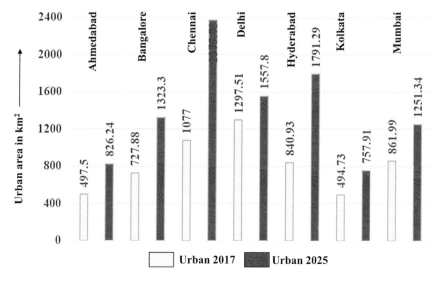

FIGURE 5.11
Urban area increase by the year 2025 for major Indian cities.

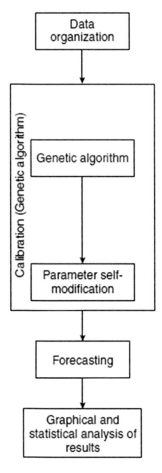

FIGURE 5.12
Genetic algorithm and SLEUTH model calibration approach. (Proposed approach based on Goldstein, 2004.)

Overall GA analysis involves calibration of metric values, number of iterations needed for calibration and assessing forecasting behaviour.

5.9 Limitations/Drawbacks of CA models

While most of the studies conducted on CA models have reported successful modelling results, several investigations also show limitations of the model. The model is highly dependent on the data quality and spatial resolution. CA fails to recognise new urban development and growth in multiple directions. The model captures and controls only spatial dynamics aspects, whereas temporal aspects can be dealt with by applying an additional model such as Markov (Eastman, 2009). CA does not have any calibration procedure and lacks the ability to account for the actual amount of change.

5.10 Concluding Remarks

LULCC is driven by anthropogenic activities that have a direct effect on the local ecosystem and the environment. LULC changes often extend beyond the city limits and create growth of urban pockets or urban sprawl, and the region lacks appropriate infrastructures and basic amenities. Modelling and visualising these changes have always been regarded, and help in formulating effective local, regional and national level policies with the sustainable management strategies. There have been significant efforts towards the development and validation of LULC models globally during the past two decades.

The chapter describes a wide range of perspectives towards modelling drivers responsible for change, scenario-based evaluation using models, various approaches available and subsequent refinements of LULC models, CA-based models and the SLEUTH model. In particular the origins underlying theory and application of the SLEUTH model and its strength are explained through integration with multiple tools, a multi-disciplinary approach and recent advanced techniques such as the genetic algorithm. However, there cannot be a single methodological approach to provide a solution for research questions of a particular region. An insightful comparison of various models on the same scale can help in developing value addition to the simulation process with the help of common measurement metrics. A case study of Ahmedabad demonstrates the process of modelling from preparation of inputs to predicted outputs. Yet there are research gaps in the validation and accuracy of models. Adopting optimisation techniques such as particle swarm optimisation, artificial intelligence and machine-learning techniques along with agent-based models seems to be the governing path of future efforts.

References

Barredo, J. I., Kasanko, M., McCormick, N., & Lavalle, C. 2003. Modelling dynamic spatial processes: Simulation of urban future scenarios through cellular automata. *Landscape and Urban Planning*, 64(3), 145–160.

Batty, M., 2009. Urban modeling. In N. Thrift, R. Kitchin (Eds.), *International encyclopedia of human geography* (pp. 51–58). Oxford, UK: Elsevier.

Batty, M., & Xie, Y. 1994. From cells to cities. *Environment and Planning B, 21*, 31–48.

Bharath, H. A., Vinay, S., Durgappa, S., & Ramachandra, T. V. 2013. Modeling and simulation of urbanisation in greater Bangalore, India. In *Proceedings of National Spatial Data Infrastructure 2013 Conference*, IIT Bombay.

Bharath, H. A., Vinay, S., & Ramachandra, T. V. 2014. Landscape dynamics modeling through integrated Markov, Fuzzy-AHP and cellular automata. In *International Geoscience and Remote Sensing Symposium*, Quebec, Canada.

Chandan, M. C., Nimish, G., & Bharath, H. A. 2019. Analysing spatial patterns and trend of future urban expansion using SLEUTH. *Spatial Information Research, 10*(3), 1–13.

Chaudhuri, G., & Clarke, K. C. 2013. The SLEUTH land use change model: A review. *The International Jorunal of Environmental Resources Research, 1*(1), 88–104.

Chen, Y. M., Li, S. Y., Li, X., & Liu, X. 2010. Simulating compact urban form using cellular automata and multicriteria evaluation: A case study in Dongguan. *Acta Scientiarum Naturalium Universitatis Sunyatseni, 49*(6), 110–114.

Clarke, K. C., & Gaydos, L. J. 1998. Loose-coupling a cellular automaton model and GIS: Long-term urban growth prediction for San Francisco and Washington/Baltimore. *International Journal of Geographical Information Science, 12*(7), 699–714.

Clarke, K. C., Hoppen, S., & Gaydos, L. 1997. A self-modifying cellular automaton model of historical urbanization in the San Francisco Bay area. *Environment and Planning B: Planning and Design, 24*(2), 247–261.

Clarke-Lauer, M. D., & Clarke, K. C. 2011. Evolving simulation modeling: Calibrating SLEUTH using a genetic algorithm. In *Proceedings of the 11th International Conference on Geo Computation* (pp. 25–29).

Couclelis, H. 1985. Cellular worlds: A framework for modeling micro–macro dynamics. *Environment and Planning A, 17*, 585–596.

Couclelis, H. 1997. From cellular automata to urban models: New principles for model development and implementation. *Environment and Planning B, 24*, 165–174.

Dietzel, C., & Clarke, K. C. 2004. Replication of spatio-temporal land use patterns at three levels of aggregation by an urban cellular automata. *Lncs, 3305*, 523–532.

Dietzel, C., & Clarke, K. C. 2007. Toward optimal calibration of the SLEUTH land use change model. *Transactions in GIS, 11*(1), 29–45.

Eastman, J. R. 2009. *IDRISI 16: The Andes edition*. Worcester: Clark University.

Ermentrout, G. B., & Edelstein-Keshet, L. 1993. Cellular automata approaches to biological modeling. *Journal of Theoretical Biology, 160*(1), 97–133.

Gardner, M. 1970. Mathematical games: The fantastic combinations of John Conway's new solitaire game "life." *Scientific American, 223*(10), 120–123.

Goldstein, N. C. 2004. Brains versus brawn-comparative strategies for the calibration of a cellular automata-based urban growth model. In P. Atkinson, G. Foody, S. Darby, F. Wu (Eds.), *GeoDynamics* (pp. 249–272). Boca Raton, FL: CRC Press.

Guan, Q., & Clarke, K. C. 2010. A general-purpose parallel raster processing programming library test application using a geographic cellular automata model. *International Journal of Geographical Information Science, 24*(5), 695–722.

He, C., Okada, N., Zhang, Q., Shi, P., & Zhang, J. 2006. Modeling urban expansion scenarios by coupling cellular automata model and system dynamic model in Beijing, China. *Applied Geography, 26*(3–4), 323–345.

Itami, R. M. 1994. Simulating spatial dynamics: Cellular automata theory. *Landscape and Urban Planning, 30,* 24–47.

Jahan, M. V., & Khosrojerdi, F. 2016. Text encryption based on glider in the game of life. *International Journal of Information Science, 6*(1), 20–27.

Jantz, C. A., Goetz, S. J., & Shelley, M. K. 2004. Using the SLEUTH urban growth model to simulate the impacts of future policy scenarios on urban land use in the Baltimore-Washington metropolitan area. *Environment and Planning B: Planning and Design, 31*(2), 251–271.

Kier, L. B., Seybold, P. G., & Cheng, C. K. 2005. *Modeling chemical systems using cellular automata* (Vol. 1). Springer Science & Business Media.

Li, X., & Gong, P. 2016. Urban growth models: Progress and perspective. *Science Bulletin, 61*(21), 1637–1650.

Li, X., & Liu, X. 2006. An extended cellular automaton using case-based reasoning for simulating urban development in a large complex region. *International Journal of Geographical Information Science, 20*(10), 1109–1136.

Lu, R. C., Huang, X. J., Zuo, T. H., Xiao, S. S., Zhao, X. F., et al. 2009. Land use scenarios simulation based on CLUE-S and Markov Composite Model—A case study of Taihu Lake Rim in Jiangsu Province. *Scientia Geographica Sinica, 29,* 577–581.

Mahiny, A. S., & Clarke, K. C. 2012. Guiding SLEUTH land-use/land-cover change modeling using multicriteria evaluation: Towards dynamic sustainable land-use planning. *Environment and Planning B: Planning and Design, 39*(5), 925–944.

Parker, D. C., Manson, S. M., Janssen, M. A., Hoffmann, M. J., & Deadman, P. 2003. Multi-agent systems for the simulation of land-use and land-cover change: A review. *Annals of the Association of American Geographers, 93*(2), 314–337.

Rafiee, R., Mahiny, A. S., Khorasani, N., Darvishsefat, A. A., & Danekar, A. 2009. Simulating urban growth in Mashad City, Iran through the SLEUTH model (UGM). *Cities, 26*(1), 19–26.

Saaty, T. L. 1980. *The analytical hierarchy process: Planning, priority setting, resource allocation.* New York: McGraw-Hill Publication.

Saaty, T. L., & Vargas, L. G. 2001. The seven pillars of the analytic hierarchy process. *International Series in Operations Research & Management Science, 34,* 27–46.

Sakieh, Y., & Salmanmahiny, A. 2016. Treating a cancerous landscape: Implications from medical sciences for urban and landscape planning in a developing region. *Habitat International, 55,* 180–191.

Sante, I., Garcia, A. M., Miranda, D., & Crecente, R. 2010. Cellular automata models for the simulation of real-world urban processes: A review and analysis. *Landscape and Urban Planning, 96*(2), 108–122.

Silva, E. A., & Clarke, K. C. 2002. Calibration of the SLEUTH urban growth model for Lisbon and Porto, Portugal. *Computers, Environment and Urban Systems, 26*(6), 525–552.

Silva, E. A., & Clarke, K. C. 2005. Complexity, emergence and cellular urban models: Lessons learned from applying Sleuth to two Portuguese metropolitan areas. *European Planning Studies, 13*(1), 93–116.

Sohl, T. L., & Sleeter, B. M. 2012. Land-use and land-cover scenarios and spatial modeling at the regional scale. *US Geological Survey Fact Sheet, 3091,* 1–4.

Sun, S., Parker, D. C., Huang, Q., Filatova, T., Robinson, D. T., et al. 2014. Market impacts on land-use change: An agent-based experiment. *Annals of the Association of American Geographers, 104*(3), 460–484.

Tang, J., Wang, L., & Yao, Z. 2007. Spatio-temporal urban landscape change analysis using the Markov chain model and a modified genetic algorithm. *International Journal of Remote Sensing*, *28*(15), 3255–3271.

Thapa, R. B., & Murayama, Y. 2009. Examining spatiotemporal urbanization patterns in Kathmandu Valley, Nepal using remote sensing and spatial metrics approaches. *Remote Sensing*, *1*, 534–556.

Tobler, W. R. 1979. Cellular Geography. In S. Gale, G. Olssen (Eds.), *Philosophy in geography* (pp. 379–386). Dordrecht, the Netherlands: D. Reidel.

Torrens, P. M. 2000. How cellular models of urban systems work. *Casa*, *160*(955), 68. Accessed on May 12, 2018. Retrieved from http://www.bartlett.ucl.ac.uk

Ulam, S. 1958. John Von Neumann 1903–1957. *Bulletin of the American Mathematical Society*, *64*(3), 1–49.

U.S. EPA. 2000. *Projecting land-use change: A summary of models for assessing the land-use patterns*. EPA/600/R-00/098. U.S. Environmental Protection Agency, Office of Research and Development, Cincinnati, OH., (September), 260 pages.

Verburg, P. H., Kok, K., Pontius, R. G., & Veldkamp, A. 2006. Modeling land-use and land-cover change. In E. Lambin, H. Geist (Eds.), *Land-use and land-cover change* (pp. 117–135). Berlin and Heidelberg: Springer.

Verburg, P. H., Veldkamp, A., Koning, G. H. J., Kok, K., & Bouma, J., 1999. A spatial explicit allocation procedure for modelling the pattern of land use change based upon actual land use. *Ecological Modelling*, *116*, 45–61.

VNRR, Voluntary National Review Report, GOI. 2017. *Report on the implementation of sustainable development goals*. Presented to the high-level political forum on sustainable development, New York. Accessed on November 8, 2018. Retrieved from http://niti.gov.in/

Wahyudi, A., & Liu, Y. 2013. Cellular automata for urban growth modeling: A chronological review on factors in transition rules. In *13th International Conference on Computers in Urban Planning and Urban Management*.

Ward, D. P., Murray, A. T., & Phinn, S. R. 2000. A stochastically constrained cellular model of urban growth. *Computer Environment Urban Systems*, *24*, 539–558.

White, R., & Engelen, G., 1994. Cellular dynamics and GIS: Modelling spatial complexity. *Geographical Systems*, *1*, 237–253.

Wolfram, S. 1986. Theory and Applications of Cellular Automata. *Advanced Series on Complex Systems Singapore World Scientific Publication 1986, 43*(12), 560.

Wolfram, S. 2002. *A new kind of science*. Champaign: Wolfram Media.

Wu, F. 1998. Simulating urban encroachment on rural land with fuzzy-logic-controlled cellular automata in a geographical information system. *Journal of Environmental Management, 53*, 293 -308.

Wu, N., & Silva, E. A. 2010. Artificial intelligence solutions for urban land dynamics: A review. *Journal of Planning Literature, 24*(3), 246–265.

Wu, X., Hu, Y., He, H. S., Bu, R., Onsted, J., et al. 2009. Performance evaluation of the SLEUTH model in the Shenyang metropolitan area of Northeastern China. *Environmental Modeling and Assessment, 14*(2), 221–230.

Yeh, A. G. O., & Li, X. 1998. Sustainable land development model for rapid growth areas using GIS. *International Journal of Geographical Information Science, 12*(2), 169–189.

Zadeh, L. A. 1965. Fuzzy sets. *Information and Control, 8*(3), 338–353.

6

Current Trends in Estimation of Land Surface Temperature Using Passive Remote Sensing Data

CONTENTS

6.1   Introduction 107
6.2   Land Surface Temperature: Theoretical Background 109
6.3   Role of Remote Sensing 114
6.4   Thermal Remote Sensors 115
6.5   LST Retrieval Algorithms 119
6.6   Conclusion 121
References 122

6.1 Introduction

Climate change was initially an academic exercise of global research-ers focused on understanding the dynamics of climatic parameters. Consequences of changes in the form of floods and droughts, and the impact on food and water security have necessitated understanding the dynamics to adopt mitigation strategies. Availability of temporal data from space-borne sensors in thermal wavelength has equipped researchers to get insights of temperature dynamics with agents of change (such as land use/land cover changes). One of the parameters that would be useful to comprehend changes in the climate is land surface temperature (LST). LST refers to the radia-tive skin temperature of Earth's surface measured from the remote sensors (Copernicus, 2018; Ese Sentinel Online, 2018) and is the basic determinant of thermal behaviour that drives the effective sweltering temperature of Earth. Factors that affect the estimation of LST include surface albedo, land surface emissivity, vegetation cover and moisture present in soil. It has been estab-lished that LST affects natural environmental cycles (water cycle, nitrogen cycle, carbon cycle, bio-geo-chemical cycle, etc.), energy exchanges in atmo-sphere and surface, wind patterns, crop patterns and cycles, biodiversity, precipitation rates, ecology, etc. (Bharath et al., 2013; Jin et al., 2015). It has been an important variable for environmental models such as weather prediction, global climatic capriciousness and ocean circulation models (Valor & Caselles, 1996; Dash et al., 2002). LST has been playing a vital role in urban climatic studies, climatology, urban heat islands, hydrological modelling, disaster management, etc. (Schmugge & Becker, 1991; Running et al., 1994; Anderson et al., 2008; Khandelwal et al., 2018; Nimish et al., 2018).

Population growth has been on the rise and 55% of the global population resides in urban areas, which is expected to rise to 68% by 2050 (UN DESA, 2018) and 90% of this increase will be in Asia and Africa. The provision of basic necessity of residence and infrastructure to cater the growing demand of the burgeoning population has resulted in large-scale land cover changes.

Urban growth and sprawl are considered primary factors altering the structure of a landscape in evolution of regional sites. The changes in landscape structure due to alteration in the spatial patterns of land use affect the evapotranspiration rates, changing the latent and sensible heat patterns (Mojolaoluwa et al., 2018). Climatic conditions at micro, meso, and macro levels are changing as a result of increasing pressure of a burgeoning population with construction activities and energy demand (domestic, transport, electricity, etc.). All these anthropogenic activities have been contributing to the escalation in the concentration of pollutants and greenhouse gases (as discussed in Chapter 2) leading to an increased skin temperature of the earth and associated thermal discomforts (Ministry of Statistic and Programme Implementation, 2015). Rapid and unplanned growth in urban structures gives rise to a phenomenon known as urban heat island (UHI) (Landsberg, 1981). UHI can be defined as the change in temperature of an urban area with respect to its surrounding rural area, as illustrated in Figure 6.1.

UHI modifies the heat and energy budget and radiation for urban landscapes. Causal factors of UHI include

- Multiple internal reflections and increased absorption of shortwave electromagnetic radiations
- Increased human-persuaded heat sources
- Obstruction in transmitted long-wave radiations
- Low evapotranspiration rate and altered heat absorption by materials used in urban areas
- Changes in wind gusts that lead to reduction in turbulent heat flux
- Reduced radiative cooling
- High heat storing (absorbing) capacity of the construction materials

The main objective of this chapter is to provide detailed information regarding land surface temperature and its estimation by remotely sensed data using various algorithms.

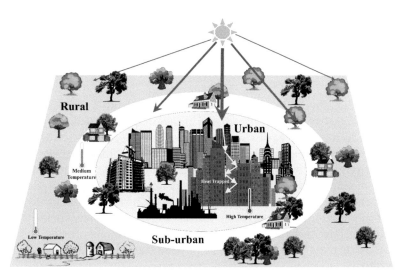

FIGURE 6.1
Temperature distribution amongst urban, suburban and rural areas signifying UHI.

6.2 Land Surface Temperature: Theoretical Background

Earth's surface energy budget is composed of radiative and non-radiative constituents. Incoming short-wave and outgoing long-wave radiations in the electromagnetic (EM) spectrum are the radiative components, while sensible and latent heat are the non-radiative components (Oke, 1982; Santra, 2019). This surface budget can be expressed as

$$Q_{\text{Total}} = S_{\text{in}} + S_{\text{ref}} + L_{\text{in}} + L_{\text{ref}} \qquad (6.1)$$

where Q_{Total} is the sum of down-welling and up-welling radiation and can be termed as net energy flux; S_{in} is the incoming short-wave radiations (diffused or direct) from Earth's surface; L_{in} is the incoming long-wave radiation from the atmosphere (clouds); S_{ref} is the reflected short-wave radiation from Earth's surface; and L_{ref} is the emitted and reflected long-wave radiation from Earth's surface.

Net energy is a sum of downward radiation that directly heats the surface of Earth and upward radiation is a combination of re-radiated energy from clouds and reflected radiation from Earth's surface. This energy budget plays a vital role in determining the temperature and microclimate of the region. One of the most important parameters that define the land surface temperature is the albedo (Oke & Cleugh, 1987), which is the ratio

of energy reflected from a surface-to-energy incident on it and can be estimated as

$$\alpha = \frac{E_{ref}}{E_T} \qquad\qquad (6.2)$$

where α is the albedo, E_{ref} is the reflected energy from a surface, and E_T is
the total energy (reflected + absorbed + transmitted) incident on the surface.

The albedo not only varies with the surface properties such as material
type, shape, colour and size but also with angle of incidence, distribution of
solar energy, cloud cover and amount of aerosols (Wang et al., 2015). Taking
example of colours, white-coloured objects have a tendency of reflecting a
major part of the incident energy, whereas black has the tendency to absorb
most of the incident energy. While in the case of materials, surfaces made
out of clay, slate and radiant barrier will have low absorption, whereas
surfaces with metal, tin or steel will have higher absorption, hence more
temperature.

Another important parameter that influences infrared thermal wave
length is emissivity and can be elaborated as the energy radiated by any
material to the energy emitted by a black body at the same atmospheric
conditions and wavelength with the same viewing angle. Emissivity being
a basic property of material depends on the composition of the material,
roughness of surface and viewing angle (Li et al., 2013). The Stephan–
Boltzmann law elucidates the relationship between energy radiated and
temperature of an object:

$$E = \sigma T^4 \qquad\qquad (6.3)$$

where E is the energy emitted, σ is the Stephan–Boltzmann constant, and T
is the temperature in kelvins.

This signifies that every object that has temperature more than 0 kelvins
radiates/emits the energy it has absorbed equivalent to the fourth power
of its absolute temperature (CSI, 2016). This is the basic principle on which
infrared/thermal sensors work. Emissivity and albedo for a few common
urban features are shown in Table 6.1.

Attempts have been made in recent times to extract the accurate surface
emissivity of various heterogeneous objects on Earth's surface using algorithms considering remote sensing data such as the normalised difference
vegetation index (NDVI)–threshold method, classification-based method,
vegetation cover method, adjusted normalised emissivity method, temperature independent spectral indices (TISI) method, and temperature emissivity separation (TES) method (Coll et al., 2001).

There are three modes of heat transfer, namely conduction, convection and
radiation. The sun's energy heats the surface of Earth directly and it can be
referred to as heat transfer due to conduction. This type of heating is high in

TABLE 6.1

Albedo and Emissivity of Common Urban Materials

Material	Albedo (α)*	Emissivity (ε)†
Asphalt	New = 0.04–0.05 Aged = 0.1–0.12 White shingle = 0.2	0.93
Concrete	New with white Portland cement = 0.7–0.8 New = 0.4–0.55 Aged = 0.2–0.3	Concrete = 0.85 Concrete tiles = 0.63 Rough concrete = 0.94
Urban materials	Galvanised steel = 0.24 Terracotta tile = 0.28 Tar and gravel = 0.33 Alabaster = 0.92 Plaster = 0.4–0.45	Red brick = 0.93 Cement = 0.54 Paint = 0.8–0.97 Mortar = 0.87 Plaster = 0.98 PVC = 0.91–0.93 Tile = 0.97
Soil (various moisture levels)	Bare = 0.17 Dark and wet = 0.05 Light and grey = 0.4 Sand = 0.15–0.45	Sand = 0.76 Soil = 0.93–0.97
Vegetation (long and short)	Green grass = 0.25 Trees = 0.15–0.2	0.9–0.99
Waterbodies (lakes, rivers)	0.03–1	0.95–0.995

*From Corona (2017).

†TET (2019) and Landsat 8 LST Analysis (2016).

the urban context due to the presence of a few materials that absorb a significant amount of heat such as asphalt and concrete. Due to the increased temperature of Earth's surface, air close to the surface gets heated up and rises due to buoyancy. Energy is transferred from one atom to another from collision. After reaching high altitudes, warm air begins to cool and sinks down and as it reaches close to the surface it again starts heating and the cycle continues. The movement of this wind and transferring energy in the form of heat can be inferred as convection heat transfer. Transfer of energy from elec tromagnetic radiations (infrared) can be referred to as radiative heat transfer and is the most important phenomenon when looked at from the perspective of satellite thermal sensing (CSI, 2016; NCCO, 2019; Lumen, 2019). Other than these three there is another mode of heat transfer that determines the temperature of Earth: advection. Movement of meteorological fronts (warm, cold, occluded, stationery) and ocean currents are examples of advection. In case of fronts, cold or warm air masses move horizontally by surface winds and interact with the system exchanging energy (Sciencing, 2019).

Planck's function is the basis of deriving LST using remote sensors. The function fundamentally relates radiative energy of a non-black or black body to its temperature. However, practically there exists no black body and for all other objects emissivity lies between 0 and 1 ($\neq$ 0 or 1). Mathematically it can be expressed as

$$R_{\lambda,T} = \varepsilon_\lambda \times B_{\lambda,T} = \varepsilon_\lambda \times \frac{C_1 \times \lambda^{-5}}{\pi \left[e^{\frac{C_2}{\lambda T}} - 1 \right]} \tag{6.4}$$

where $R_{\lambda,T}$ is the spectral radiance of the non-black body at wavelength and temperature T (in Wm^{-2} μm^{-1} sr^{-1}); $B_{\lambda,T}$ is the spectral radiance of black body at wavelength and temperature T (in Wm^{-2} μm^{-1} sr^{-1}) (Planck's radiation law); is the wavelength (μm); T is the temperature (K); ε is the emissivity of the body at wavelength ; and C_1 and C_2 are radiation constants.

As per Kirchhoff's law of radiation, a good absorber will be a good emitter. This holds good for systems at local thermodynamic equilibrium signifying the system having single temperature thermodynamically. A 50–70 km range of atmosphere from the surface of Earth can be anticipated following this condition (Dash et al., 2002). Separating the reflected downwelling irradiance and upwelling radiance, the temperature of a Lambertian surfaced object of known emissivity can be quantified by rearranging Equation 6.4 as shown in Equation 6.5:

$$T = \frac{C_2}{\lambda \ln \left(\frac{\varepsilon_\lambda C_1}{\pi \lambda^5 R_{\lambda,T}} + 1 \right)} \tag{6.5}$$

For deriving the relationship between temperature and wavelength, Wein's displacement law is used that states that the black-body radiation will peak at different temperatures for different wavelengths and are inversely proportional to each other as shown in Figure 6.1. Wein's displacement law can be mathematically written as

$$\lambda_m \times T = 2897\ \mu mK \tag{6.6}$$

where T is the temperature and λ_m is the wavelength at which maximum radiation is achieved.

The black-body radiation curve (Chapter 1 in Lillesand et al., 2003) indicates that as the temperature increases the peak intensity of radiation is achieved at lower wavelengths (black-body radiation). The mean temperature of Earth is around 300 K (27°C). Thus, as per Wein's law, the wavelength at which the earth radiates is at 9.66 μm. Thus, the thermal range for estimation of LST is from 8 to 15 μm. In this electromagnetic region the atmospheric absorption is minimum, and maximum transmission occurs with minimal attenuations, making it possible to perform thermal analysis using remote sensors. Other than the signal-to-noise ratio, another factor that plays a significant role is sensitivity between radiance and surface temperature that can be defined by first derivative of Planck's function with respect to temperature:

$$\frac{dB}{dT} = \frac{C_1 C_2 \lambda^{-6} e^{\frac{C_2}{\lambda T}}}{\pi T^2 \left(e^{\frac{C_2}{\lambda T}} - 1 \right)^2}$$

(6.7)

Considering a black body when T is kept as 300 K, the peak is obtained at around 9.7 μm wavelength. Radiative transfer models are used for estimation of this relationship between temperature and radiance, and they work under some basic assumptions: (1) there exists a local thermodynamic equilibrium that is valid up to 50–70 km, (2) a clear atmosphere with no haze is considered that infers zero scattering, and (3) Earth's surface is a Lambertian surface.

A long-wave infrared sensor measures the radiations from the objects present at the surface of Earth's atmosphere along the sensor's line of sight. Considering a cloud-free atmosphere under local thermodynamic equilibrium as shown in Figure 6.2, the radiance received at the sensor can be represented as shown in equation 6.8. This equation is termed the radiative transfer equation:

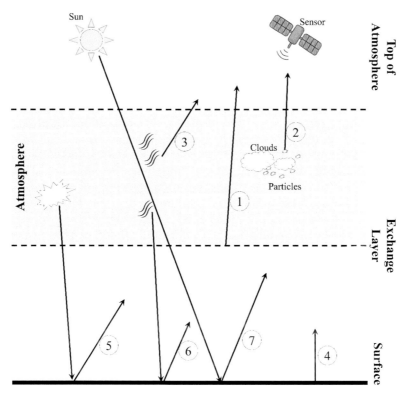

FIGURE 6.2
Illustration of radiative transfer equation in thermal infrared region.

$$L_{\text{sensor},\lambda} = \left[\varepsilon_\lambda B_\lambda \left(T_s \right) + \left(1 - \varepsilon_\lambda \right) L^{\downarrow}_{\text{atm},\lambda} \right] \tau_\lambda + L^{\uparrow}_{\text{atm},\lambda} \qquad (6.8)$$

where $L_{\lambda\ \text{sensor}}$ is the top of atmospheric radiance at wavelength λ; ε_λ is emissivity at wavelength λ; B_λ is black-body radiance at wavelength obtained by Plank's law (function of T_s); T_s is the LST; $L^{\downarrow}_{\text{atm}}$ is downwelling atmospheric radiance; τ_λ is the total atmospheric transmissivity between the surface and sensor at wavelength λ; and $L^{\uparrow}_{\text{atm}}$ is the upwelling atmospheric radiance.

Figure 6.2 shows the total radiance reaching a satellite and their sources. Path ① represents radiance detected at ground level that is attenuated due to atmospheric interaction. Path ② represents the upward thermal radiance due to atmospheric particles and clouds. Path ③ represents radiance due to upward solar diffusion. Path ④ represents the radiance emitted by the features/objects on the surface of Earth. Path ⑤ represents the downwards thermal atmospheric radiance. Path ⑥ represents solar-diffused radiance reflected by features/objects on Earth's surface. Path ⑦ represents the radiance due to reflected radiation from the surface. Here paths ①, ② and ③ can be considered under upwelling radiance; paths ⑤, ⑥ and ⑦ are downwelling radiance; and path ④ is the emissivity fraction when termed as parameters in Equation 6.8.

6.3 Role of Remote Sensing

Multi-resolution remote sensing data has wide applications in almost all the domains, disciplines and fields as a result of its ease and technological advancement. Remotely sensed data is considered as the fastest and economical method for mapping large areas and areas that are inaccessible for humans. With the availability of high-resolution data both in terms of spatial and spectral, Earth-observation-based monitoring of urban development has become one of the major applications of remote sensing (Chen et al., 2000; Ji et al., 2001). It helps urban planners to understand landscape with respect to topology, land cover and land use maps. Sensors in varied electromagnetic ranges capture heterogeneous features of Earth in a different way making it easy to distinguish multiple features. It also plays a significant role in various environmental and climate studies. Climatic, meteorological and environmental studies first started in 1960 when the first meteorological satellite was launched, and since then there has been tremendous growth in these fields (Tomlinson et al., 2011). Researchers have developed algorithms to provide better, accurate and precise spatial data, compared to point data available from meteorological stations at regional as well as at global levels, thus enhancing the quality of information (Jin & Shepherd, 2005; Mendelsohn et al., 2007). Incorporation of sensors in the thermal infrared range (8–15 μm) has offered the possibility of measuring LST at a global scale spatially at regular intervals compared to ground-based measurements that are too tiresome and erroneous. Availability of spatial data at regular intervals has facilitated development of multiple algorithms across the globe for estimation of LST from the spatial thermal data acquired through the geostationary and polar satellites. Detailed descriptions about thermal sensors and various algorithms are discussed in Sections 6.4 and 6.5, respectively.

6.4 Thermal Remote Sensors

There has been development in thermal remote sensing during the current century and there are numerous thermal satellite sensors providing thermal-based spatio-temporal information of the Earth's surface. Climatologists and meteorologists have data acquired through various remote sensing platforms for calculating LST. The first ever thermal data was collected in 1960 by the US satellite TIROS (Television Infrared Operational Satellite) for monitoring frontal movement and regional cloud patterns (Wark et al., 1962; Jensen, 2009). Subsequently, NASA (National Aeronautics and Space Administration) launched HCMM (Heat Capacity Mapping Mission) in 1978 that provided daily thermal infrared data (twice) at a spatial resolution of 600 m. Later, NASA launched multiple satellites for ocean temperature mapping, monitoring sea surface temperature, acquiring terrestrial thermal energy, etc. During the 20th century, Landsat 4 and 5 (Thematic Mapper) acquired thermal data between 10.4 and 12.5 µm at a spatial resolution of 120 m. At present there are multiple satellites such as Landsat 7 (Enhanced Thematic Mapper Plus), Landsat 8 (Operational Land Imager/Thermal Infrared Sensors), AVHRR (Advanced Very High Resolution Radiometer), MODIS (Moderate Resolution Imaging Spectro-Radiometer), NOAA GOES (Geostationary Operational Environmental Satellite) and ASTER (Advanced Spaceborne Thermal Emission and Reflection Radiometer) offering global multi-resolution thermal data (Pu et al., 2006). Now, sophisticated airborne sensors such as ATLAS and AHS are also available with the technological development that provides thermal data at finer resolution for micro-level studies (Gluch et al., 2006; Sobrino et al., 2006). Tables 6.2 and 6.3 show the timeline and information on the data from thermal sensors.

1. *GOES*: The Geostationary Operational Environmental Satellite is a system operated by the United States' NOAA (National Oceanic and Atmospheric Administration) for weather forecasting, severe storm monitoring and research related to meteorology. It carries a multi-spectral sensor having two thermal channels: 10.2–11.2 µm and 11.5–12.5 µm with a spatial resolution of 4 km. It can be

TABLE 6.2

Data Availability of Satellites with Thermal Sensors

Satellite/Sensor	Data Availability
Landsat 4	1983–1994
Landsat 5	1984–2011
GOES	1994–present
Landsat 7	1999–2013
MODIS Terra	2000–present
ASTER	2000–present
MODIS Aqua	2002–present
MetOP AVHRR	2007–present
Landsat 8	2013–present

TABLE 6.3

Information about Current Satellites and Sensors Capable of Capturing Thermal Data

Satellite/Sensor	Spatial Resolution	TIR Spectral Bands	Temporal Resolution
GOES	4 km	Band 4: 10.2–11.2 μm Band 5: 11.5–12.5 μm	Geostationary
MODIS Aqua	1 km	Band 31: 10.78–11.28 μm Band 32: 11.77–12.27 μm	Twice daily
MODIS Terra	1 km	Band 31: 10.78–11.28μm Band 32: 11.77–12.27 μm	Twice daily
Landsat TM	120 m*	Band 6: 10.4–12.5 μm	16 days
Landsat ETM+	60 m*	Band 6: 10.4–12.5 μm	16 days
Landsat OLI/TIRS	100 m*	Band 10: 10.5–11.19 μm Band 11: 11.50–12.51 μm	16 days
AVHRR MetOP	1.1 km	Band 4: 10.3–11.3 μm Band 5: 11.5–12.5 μm	29 days
AVHRR NOAA	1.1 km	Band 4: 10.3–11.3 μm Band 5: 11.5–12.5 μm	Twice daily
ASTER Terra	90 m	Band 10: 8.125–8.475 μm Band 11: 8.475–8.825 μm Band 12: 8.925–9.275 μm Band 13: 10.25–10.95 μm Band 14: 10.95–11.65 μm	Twice daily
AATSR Envisat	1 km	Band 11 Band 12	35 days
SEVIRI Meteosat-8	3 km	Band 4: 10.2–11.2 μm Band 5: 11.5–12.5 μm	Geostationary

*Collected at provided resolution and then resampled to 30 m as final product for users.

used to perform measurements related to LST using a single channel or multiple channels (Sun et al., 2004). Sun and Pinker (2003) estimated LST by developing two algorithms using GEOS-8 data. Inamdar et al. (2008) through MODIS data calibrated GOES data, that provided a 1 km LST data set with a temporal resolution of half an hour and accuracy less than 2°C.

2. *MODIS*: The Moderate Resolution Imaging Spectro-Radiometer (MODIS) is a payload imaging sensor that was launched by NASA on two satellites: Terra (EOS AM) and Aqua (EOS PM). It is a hyperspectral sensor with 36 bands of resolutions at 250 m, 500 m and 1 km. It captures the same object on Earth twice every day. MODIS has daily LST products available that are freely available for users across the globe. These products (MYD11A1 and MOD11A1) are derived from a split window algorithm that takes into account band 31 (10.78–11.28 μm) and band 32 (11.77–12.27 μm). These products have been verified and found of relatively good accuracies of 1 K over homogeneous surfaces (Wan et al., 2004; Coll et al., 2005; Wan, 2008). Numerous studies on urban climatology utilise MODIS LST products. Zhou et al. (2014) estimated LST by integrating regression models and genetic algorithms keeping the data source as MODIS products. Azevedo et al. (2016) used MODIS data to quantify the day- and night-time urban heat island (UHI) in Birmingham, UK, and compared it with high-resolution air temperature observations. Similarly, Yanev and Filchev (2016) evaluated a MODIS level-3 LST product with in situ temperature data for understanding the UHI phenomenon across Sofia. An interesting comparison of Landsat data and MODIS was performed in a study to estimate water temperature over Lake Titicaca, and it was found that both satellite images have comparable accuracies with respect to LST (Ruiz-Verdu et al., 2016).

3. *LANDSAT*: This is one of the most popular and reliable data sources since 1973. Thermal sensors were introduced from Landsat series 4 onwards (since 1983). Landsat 4 and 5 were tailored with a sensor called thematic mapper that included seven bands (three visible, one near infrared, two short-wave infrared and one thermal infrared). The thermal infrared band (10.4–12.5 μm) data at a spatial resolution of 120 m (with temporal resolution of 16 days) are resampled to 30 m before making it available for users. After the success and popularity of this satellite, another satellite was launched, but it did not enter its orbit and turned out to be first failed mission of the Landsat series. Afterwards in 1999, Enhanced Thematic Mapper Plus (ETM+) with an additional panchromatic band was launched on Landsat 7. This satellite collected thermal information at 60 m, which was resampled to 30 m. But due to some technical failure, in 2003 the sensor

started providing data with a scan line error that covered about 20% of the total scene. The excellent quality of data and reduced pricing led to an increased number of users and as a result, USGS made data available in public domain in October 2008, increasing downloads by almost 60 times. In 2013, the latest Landsat was launched with 11 bands including 2 thermal bands that collected thermal data at a resolution of 100 m and resampled it to 30 m before providing it to the user. One big disadvantage of Landsat is that it does not collect night-time data and does not allow day–night variation of temperature. Sobrino et al. (2004) and Li et al. (2004) developed a technique to retrieve land surface temperature from Landsat 5. This data was used to perform UHI study for Dallas, Texas, using extracted tree cover and thermal sensors (Aniello et al., 1995). Kuscu and Sengezer (2011) determined heat islands with Landsat Thematic Mapper data and found a relationship between surface temperature and urbanisation for Istanbul. Grover and Singh (2016) performed a similar study for the Indian city of Mumbai. Similarly, a number of studies have been performed using Landsat 7 (ETM+) data. Stathopoulou and Cartalis (2007) calculated the heat islands and related it with land cover data. Mallick et al. (2008) estimated LST over Delhi using the Landsat ETM sensor. Kumar et al. (2012) derived UHIs by estimating LST for Vijaywada city, Andhra Pradesh, India. Rajasekar and Weng (2009) studied UHIs using Landsat 5 and 7 and derived the variations in temperature across the years. In the case of Landsat 8, due to the presence of two thermal bands in different wavelength regions, algorithms such as split-window and multi-channel were used to derive LST. Du et al. (2015) and Anandababu et al. (2018) estimated LST using the split-window algorithm for Beijing and Hosur, India, respectively. Tsou et al. (2017) assessed UHI for Shenzhen and Hong Kong using Landsat 8 data.

4. *AVHRR*: The Advance Very High Resolution Radiometer serves as a sensor on a sun-synchronous NOAA satellite. This sensor offers a spatial resolution of 1.1 km with daily coverage. It has two channels sensitive to thermal radiations: band 4 (10.3–11.3 μm) and band 5 (11.5–12.5 μm). Other than NOAA few other platforms such as MetOP (polar orbiting satellite) have this sensor on-board. This sensor offers large historical data and provides a data set once every month. Vogt (1996) retrieved LST from NOAA AVHRR data. One study mapped micro-UHI using NOAA and CORINE land cover for Greece (Stathopoulou et al., 2004). One of the biggest disadvantages of AVHRR is its inability to provide night-time images.

5. *ASTER*: The Advanced Spaceborne Thermal Emission and Reflection Radiometer (ASTER) is the only satellite with more than one or two bands for thermal. ASTER has five bands in the thermal EM range:

band 10 (8.125–8.475 μm), band 11 (8.475–8.825 μm), band 12 (8.925–9.275 μm), band 13 (10.250–10.950 μm) and band 14 (10.950–11.650 μm) (Yamaguchi et al., 1998). This sensor captures data twice diurnally and provides it at a resolution of 90 m. This sensor acquires data only on request and is not free of cost. This satellite has an advantage of fine spatial and temporal resolution that no other satellite provides. The AST08 product by ASTER calculates temperature using the temperature–emissivity separation (TES) method. ASTER products and images have been used for numerous environmental-related applications. One such interesting study was performed by Kato and Yamaguchi (2005). They analysed the UHI effect using ASTER and Landsat ETM+, and separated anthropogenic and natural heat radiations. Cai et al. (2011) monitored the UHI effect in Beijing by combining ASTER and Landsat TM data. Weng et al. (2011) modelled UHI and developed a relationship with impervious surface as well as fraction of vegetation cover (Table 6.3).

6.5 LST Retrieval Algorithms

Ambient air temperature measured at meteorological stations provides only point data that is usually extrapolated for the entire region with inaccuracies. This problem is overcome by using remote sensing data, as it provides pixel-by-pixel values for surface temperature. LST is not just measured directly from airborne sensors; instead the thermal sensor captures the at-satellite brightness temperature that is equivalent to the amount of long-wave energy radiated by various features/objects on Earth's surface. Later, emissivity and other atmospheric corrections are incorporated to provide LST (Li & Becker, 1993; Ghent et al., 2017). The relationship between LST and ambient air temperature was derived to overcome the inaccuracy problem due to point measurement (Shen & Leptoukh, 2011). A few important factors that play an integrating role in defining and estimating LST are cloud cover, sensor calibration, atmospheric corrections, emissivity and effective parameter definition. Various computational algorithms have been developed related to LST that capture strong spatial, temporal and spectral variations, but they are quite complex and require multiple parameters (Labed & Stoll, 1991; Salisbury & D'Aria, 1992). A few of the algorithms for retrieval of LST are as follows:

1. Single-channel/single-window algorithm: This is one of the most common methods used to estimate LST, and its uses a single band to quantify LST. It is a three-step process: (1) conversion of digital number into radiance, (2) conversion of radiance into at-satellite brightness temperature and (3) conversion of at-satellite brightness

temperature into LST by incorporating the emissivity of each feature (Xiao & Weng, 2007; Jimenez-Munoz et al., 2009; Qin et al., 2001;
Nimish et al., 2018). This method's shortcoming is that it does not
consider the atmospheric effect and water vapour. An improved
single-channel algorithm was developed by Yu et al. (2014) considering the effect of water vapour. Another improved single-channel
algorithm was derived by Cristobal et al. (2018) that considered near-
surface air temperature and atmospheric column of water vapour.
However, the atmospheric attenuation and profile variations were
not highlighted in these algorithms.

2. Split-window algorithm: This algorithm was first proposed by Anding
and Kauth (1970) and improved by McMillin (1971). It was named split-
window as it takes two thermal channels into consideration and utilises the difference in atmospheric absorption in two thermal channels
centred at 11 µm and 12 µm. Out of these two channels, one is more sensitive to atmospheric water vapour than other. This method basically
utilises two differentially absorbing channels in the thermal region for
removing attenuation caused by atmospheric absorption. The radiance
measured by each of the thermal channels in this method is related to
the diminution suffered by atmosphere due to surface-emitted radiance (McMillin, 1975). The accuracy of this algorithm is high, but prior
knowledge of a few parameters related to atmosphere such as emissivity and water vapour is necessary (Benmecheta et al., 2013). There are
two types of split-window algorithms as per the literature.

   a. Linear split-window algorithm: This algorithm articulates LST
   as a linear function of the at-satellite brightness temperature as
   captured by two thermal channels. It can be written as

$$LST = a_0 + a_1 T_i + a_2 (T_i - T_j)$$ (6.9)

   where a_k are coefficients that are influenced primarily by the
   spectral response of two channels, emissivity of two channels,
   water vapour and viewing zenith angle; and T_i and T_j are the at-
   satellite brightness temperatures of two channels – i and j.

   b. Nonlinear split-window algorithm: There are certain errors that
   add up when we consider a linear relationship. As a result of a
   few approximations with respect to transmittance in the linear
   split-window algorithm, there might be inaccurate values of LST
   for hot and wet regions. To overcome this, a nonlinear relationship is applied between brightness temperatures of two channels, as given in the following equation:

$$LST = a_0 + a_1 T_i + a_2 (T_i - T_j) + a_3 (T_i - T_j)^2$$ (6.10)

These coefficients are estimated by simulating the atmospheric profiles using MODTRAN and other sources. These coefficients can be parameterised considering only emissivity; emissivity and water vapour; and emissivity, water vapour and viewing zenith angle. Based on the assumptions the accuracy of estimation varies (Li et al., 2013).

3. Radiative transfer algorithm: It is a single-step process that takes into account complex mathematical equation as shown in Section 6.2. It re-formulates Equation 6.8 as

$$LST = \frac{C_1}{\lambda_i \ln \left(\dfrac{C_2}{\lambda_i^5 \left(L_{sensor,\lambda} - L^{\uparrow}_{atm,\lambda} - \tau_\lambda \left(1 - \varepsilon_\lambda\right) L^{\downarrow}_{atm,\lambda} \right) / \tau_\lambda \varepsilon_\lambda} \right)} \quad (6.11)$$

where λ_i is the effective band wavelength for band i; C_1 is 14387.7 (μm K); and C_2 is 1.19104×10^8 (μm^4 m^{-2} sr^{-1}).

Research efforts are under way to improve these basic algorithms. Sun and Pinker (2003) developed a modified split-window algorithm that takes into account inaccuracy in radiance and improves the overall retrieval. Subsequently, another algorithm utilising three channels (mid-infrared and thermal channels) aims to improve the surface temperature measurement accuracies. Sobrino et al. (2006) developed a method called temperature emissivity separation that uses NEM, RATIO and MMD modules to quantify emissivity and temperature, but one of the major drawbacks of this method is the availability of fine spatial resolution thermal data.

6.6 Conclusion

The chapter highlights the application of remotely sensed data through computation of land surface temperature (LST) with the theoretical background. It also encompasses information regarding the thermal sensors available and various algorithms that can effectively retrieve LST from the sensor data. The NOAA study (2018) highlights that the first decade of the 21st century had the warmest global temperatures recorded since the beginning of the 19th century. As per another government report, out of the ten warmest years that occurred since the 19th century, nine are from 2007–2017, 2016 being the warmest. Most cities around the world are facing the issue of increased thermal instability and reduced comfort levels. These issues are leading to increased health problems (physical as well as psychological) for residents. Measures for adapting a rise of 2°C in global temperature will cost around 70 billion–100 billion USD. This chapter helps in identifying and understanding the problems associated with higher surface temperatures. This would help in prudent decision making by policymakers, stakeholders and government officials in mitigating the effects of changes in climate with escalating global warming and counteracting the growing human needs by means of sustainable development.

References

Anandababu, D., Purushothaman, B. M., & Babu, S. S. 2018. Estimation of land surface temperature using LANDSAT 8 data. *International Journal of Advance Research, Ideas and Innovations in Technology*, 4(2), 177–186.

Anderson, M. C., Norman, J. M., Kustas, W. P., Houborg, R., Starks, P. J., et al. 2008. A thermal-based remote sensing technique for routine mapping of land-surface carbon, water and energy fluxes from field to regional scales. *Remote Sensing of Environment*, 112, 4227–4241.

Anding, D., & Kauth, R. 1970. Estimation of sea surface temperature from space. *Remote Sensing of Environment*, 1(4), 217–220.

Aniello, C., Morgan, K., Busbey, A., & Newland, L. 1995. Mapping micro-urban heat islands using Landsat TM and a GIS. *Computers & Geosciences*, 21(8), 965–969.

Azevedo, J., Chapman, L., & Muller, C. 2016. Quantifying the daytime and night-time urban heat island in Birmingham, UK: A comparison of satellite derived land surface temperature and high resolution air temperature observations. *Remote Sensing*, 8(2), 153.

Benmecheta, A., Abdellaoui, A., & Hamou, A. 2013. A comparative study of land sur-face temperature retrieval methods from remote sensing data. *Canadian Journal of Remote Sensing*, 39(1), 59–73.

Bharath, S., Rajan, K. S., & Ramachandra, T. V. 2013. Land surface temperature responses to land use land cover dynamics. *Geoinformatics Geostatistics: An Overview*, 1(4), 1–10.

Cai, G., Du, M., & Xue, Y. 2011. Monitoring of urban heat island effect in Beijing combin-ing ASTER and TM data. *International Journal of Remote Sensing*, 32(5), 1213–1232.

Chen, S. P., Zeng, S., & Xie, C. G. 2000. Remote sensing and GIS for urban growth anal-ysis in China. *Photogrammetric Engineering and Remote Sensing*, 66(5), 593–598.

Coll, C., Caselles, V., Galve, J. M., Valor, E., Niclos, R., et al. 2005. Ground measure-ments for the validation of land surface temperatures derived from AATSR and MODIS data. *Remote Sensing of Environment*, 97(3), 288–300.

Coll, C., Sobrino, J., Caselles, V., Jiménez, J. C., Rubio, E., et al. 2001. A comparison of methods for surface temperature and emissivity estimation. *The Digital Airborne Spectrometer Experiment (DAISEX)*,499, 217–223.

Copernicus. 2018. *Land surface temperature.* Accessed on May 14, 2019. Retrieved from https://land.copernicus.eu/global/products/lst

Corona. 2017. *List of reflectance/albedo of common materials.* Accessed on May 09, 2019. Retrieved from https://corona-renderer.com/

Cristóbal, J., Jiménez-Muñoz, J., Prakash, A., Mattar, C., Skoković, D., et al. 2018. An improved single-channel method to retrieve land surface temperature from the Landsat-8 thermal band. *Remote Sensing*, 10(3), 431.

CSI. 2016. *Energy: The driver of climate – Temperature and radiation.* Accessed on May 09, 2019. Retrieved from http://www.ces.fau.edu/

Dash, P., Göttsche, F. M., Olesen, F. S., & Fischer, H. 2002. Land surface temperature and emissivity estimation from passive sensor data: Theory and practice-current trends. *International Journal of Remote Sensing*, 23(13), 2563–2594.

Du, C., Ren, H., Qin, Q., Meng, J., & Zhao, S. 2015. A practical split-window algorithm for estimating land surface temperature from Landsat 8 data. *Remote Sensing*, 7(1), 647–665.

Ese Sentinel Online 2018. *Land surface temperature.* Accessed on May 14, 2019. Retrieved from https://sentinel.esa.int/

Ghent, D. J., Corlett, G. K., Göttsche, F. M., & Remedios, J. J. 2017. Global land surface temperature from the along-track scanning radiometers. *Journal of Geophysical Research: Atmospheres*, 122(22), 12167–12193.

Gluch, R., Quattrochi, D. A., & Luvall, J. C. 2006. A multi-scale approach to urban thermal analysis. *Remote Sensing of Environment*, 104(2), 123–132.

Grover, A., & Singh, R. B. 2016. Monitoring spatial patterns of land surface temperature and urban heat island for sustainable megacity: A case study of Mumbai, India, using Landsat TM data. *Environment and Urbanization ASIA*, 7(1), 38–54.

Inamdar, A. K., French, A., Hook, S., Vaughan, G., & Luckett, W. 2008. Land surface temperature retrieval at high spatial and temporal resolutions over the southwestern United States. *Journal of Geophysical Research: Atmospheres*, 113(D7), 1–10.

Jensen, J. R. 2009. *Remote sensing of the environment: An earth resource perspective*, 2nd edition. India: Pearson Education. Accessed on May 10, 2019. Retrieved from http://www.gers.uprm.edu/

Ji, C., Liu, Q., Sun, D., Wang, S., Lin, P., et al. 2001. Monitoring urban expansion with remote sensing in China. *International Journal of Remote Sensing*, 22(8), 1441–1455.

Jimenez-Munoz, J. C., Cristobal, J., Sobrino, J. A., Soria, G., Ninyerola, M., et al. 2009. Revision of the single-channel algorithm for land surface temperature retrieval from landsat thermal-infrared data. *IEEE Transactions on Geoscience and Remote Sensing*, 47(1), 339–349.

Jin, M., & Shepherd, J. M. 2005. Inclusion of urban landscape in a climate model: How can satellite data help? *Bulletin of the American Meteorological Society*, 86(5), 681–689.

Jin, M., Li, J., Wang, C., & Shang, R. 2015. A practical split-window algorithm for retrieving land surface temperature from Landsat-8 data and a case study of an urban area in China. *Remote Sensing*, 7(4), 4371–4390.

Kato, S., & Yamaguchi, Y. 2005. Analysis of urban heat-island effect using ASTER and ETM+ data: Separation of anthropogenic heat discharge and natural heat radiation from sensible heat flux. *Remote Sensing of Environment*, 99(1–2), 44–54.

Khandelwal, S., Goyal, R., Kaul, N., & Mathew, A. 2018. Assessment of land surface temperature variation due to change in elevation of area surrounding Jaipur, India. *Egyptian Journal of Remote Sensing and Space Science*, 21(1), 87–94.

Kumar, K. S., Bhaskar, P. U., & Padmakumari, K. 2012. Estimation of land surface temperature to study urban heat island effect using Landsat ETM+ image. *International Journal of Engineering Science and Technology*, 4(2), 771–778.

Kuscu, C., & Sengezer, B. April, 2011. Determination of heat islands from Landsat TM data: Relationship between surface temperature and urbanization factors in Istanbul. Paper Presented at 34th International Symposium on Remote Sensing of Environment, Sydney, Australia. Accessed on May 11, 2019. Retrieved from http://citeseerx.ist.psu.edu/viewdoc/download?doi=10.1.1.368.2094&rep= rep1&type=pdf

Labed, J., & Stoll, M. P. 1991. Spatial variability of land surface emissivity in the thermal infrared band: Spectral signature and effective surface temperature. *Remote Sensing of Environment*, *38*(1), 1–17.

Landsat 8 LST Analysis. 2016. *Automated generation of land surface temperature estimates*. Accessed on March 28, 2018. Retrieved from https://datahub.cmap.illinois.gov/

Landsberg, H. E. 1981. *The urban climate*. New York: Academic press.

Li, Z. L., & Becker, F. 1993. Feasibility of land surface temperature and emissivity determination from AVHRR data. *Remote Sensing of Environment*, *43*(1), 67–85.

Li, F., Jackson, T. J., Kustas, W. P., Schmugge, T. J., French, A. N., et al. 2004. Deriving land surface temperature from Landsat 5 and 7 during SMEX02/SMACEX. *Remote Sensing of Environment*, *92*(4), 521–534.

Li, Z. L., Tang, B. H., Wu, H., Ren, H., Yan, G., et al. 2013. Satellitederived land surface temperature: Current status and perspectives. *Remote Sensing of Environment*, *131*, 14–37.

Lillesand, T. M., Kiefer, R. W., Chipman, J. W., & Lilles, T. M. 2003. *Remote sensing and image interpretation*, 5th edition. New York: Wiley, John and Sons.

Lumen. 2019. *Heat transfer in the atmosphere*. Accessed on May 09, 2019. Retrieved from https://courses.lumenlearning.com/

Mallick, J., Kant, Y., & Bharath, B. D. 2008. Estimation of land surface temperature over Delhi using Landsat-7 ETM+. *Journal of Indian Geophysical Union*, *12*(3), 131–140.

McMillin, L. M. 1971. A method of determining surface temperatures from measurements of spectral radiance at two wavelengths (Doctral Dissertation). Accessed on April 02, 2019. Retrieved from http://citeseerx.ist.psu.edu/viewdoc/download?doi=10.1.1.1019.7897&rep=rep1&type=pdf

McMillin, L. M. 1975. Estimation of sea surface temperatures from two infrared window measurements with different absorption. *Journal of Geophysical Research*, *80*(36), 5113–5117.

Mendelsohn, R., Kurukulasuriya, P., Basist, A., Kogan, F., & Williams, C. 2007. Climate analysis with satellite versus weather station data. *Climatic Change*, *81*(1), 71–83.

Ministry of Statistics and Programme Implementation, Government of India. 2015. *Statistics related to climate change - India 2015*. Accessed on May 14, 2019. Retrieved from http://www.mospi.gov.in/sites/default/files/publication_reports/climateChangeStat2015.pdf

Mojolaoluwa, T. D., Emmanuel, O. E., & Kazeem, A. I. 2018. Assessment of thermal response of variation in land surface around an urban area. *Modelling Earth Systems and Environment*, *4*(2) 535–553.

NCCO. 2019. *Conduction*. Accessed on May 09, 2019. Retrieved from https://climate.ncsu.edu/

Nimish, G., Chandan, M. C., & Bharath, H. A. 2018. Understanding current and future landuse dynamics with land surface temperature alterations: A case study of Chandigarh. *ISPRS Annals of Photogrammetry, Remote Sensing & Spatial Information Sciences*, *4*(5), 79–86.

NOAA. 2018. *A paleo perspective on global warming*. Accessed on May 14, 2019. Retrieved from https://www.ncdc.noaa.gov/global-warming

Oke, T. R. 1982. The energetic basis of the urban heat island. *Quarterly Journal of the Royal Meteorological Society, 108*(455), 1–24.

Oke, T. R., & Cleugh, H. A. 1987. Urban heat storage derived as energy balance residuals. *Boundary-Layer Meteorology, 39*(3), 233–245.

Pu, R., Gong, P., Michishita, R., & Sasagawa, T. 2006. Assessment of multi-resolution and multi-sensor data for urban surface temperature retrieval. *Remote Sensing of Environment, 104*(2), 211–225.

Qin, Z., Karnieli, A., & Berliner, P. 2001. A mono-window algorithm for retrieving land surface temperature from Landsat TM data and its application to the Israel-Egypt border region. *International Journal of Remote Sensing, 22*(18), 3719–3746.

Rajasekar, U., & Weng, Q. 2009. Spatio-temporal modelling and analysis of urban heat islands by using Landsat TM and ETM+ imagery. *International Journal of Remote Sensing, 30*(13), 3531–3548.

Ruiz-Verdú, A., Jiménez, J. C., Lazzaro, X., Tenjo, C., Delegido, J., et al. July, 2016. *Comparison of MODIS and Landsat-8 retrievals of chlorophyll-a and water temperature over Lake Titicaca. Paper Presented at IEEE International Geoscience and Remote Sensing Symposium (IGARSS).* Accessed on May 09, 2019. Retrieved from https://ieeexplore.ieee.org/document/7730993

Running, S. W., Justice, C. O., Salomonson, V., Hall, D., Barker, J., et al. 1994. Terrestrial remote sensing science and algorithms planned for EOS/MODIS. *International Journal of Remote Sensing, 15*(17), 3587–3620.

Salisbury, J. W. & D'Aria, D. M. 1992. Emissivity of terrestrial materials in the 8–14 µm atmospheric window. *Remote Sensing of Environment, 42*(2), 83–106.

Santra, A. 2019. Land surface temperature estimation and urban heat island detection: A remote sensing perspective. In Santra, A., & Mitra, S. S. (Eds.), *Environmental information systems: Concepts, methodologies, tools, and applications* (pp. 1538–1560). Hershey, PA: IGI Global.

Schmugge, T. J., & Becker, F. 1991. Remote sensing observations for the monitoring of landsurface fluxes and water budgets. In Schmugge, T. J., & André, J. C. (Eds.), *Land surface evaporation* (pp. 337–347). New York: Springer.

Sciencing. 2019. *The difference between convection & advection heat transfers.* Accessed on May 09, 2019. Retrieved from https://sciencing.com/

Shen, S., & Leptoukh, G. G. 2011. Estimation of surface air temperature over central and eastern Eurasia from MODIS land surface temperature. *Environmental Research Letters, 6*(4), 1–8.

Sobrino, J. A., Jiménez-Muñoz, J. C., & Paolini, L. 2004. Land surface temperature retrieval from LANDSAT TM 5. *Remote Sensing of Environment, 90*(4), 434–440.

Sobrino, J. A., Jiménez-Muñoz, J. C., Zarco-Tejada, P. J., Sepulcre-Canto, G., & de Miguel, E. 2006. Land surface temperature derived from airborne hyperspectral scanner thermal infrared data. *Remote Sensing of Environment, 102*(1–2), 99–115.

Stathopoulou, M., & Cartalis, C. 2007. Daytime urban heat islands from Landsat ETM+ and Corine land cover data: An application to major cities in Greece. *Solar Energy, 81*(3), 358–368.

Stathopoulou, M., Cartalis, C., & Keramitsoglou, I. 2004. Mapping micro-urban heat islands using NOAA/AVHRR images and CORINE land cover: An application to coastal cities of Greece. *International Journal of Remote Sensing, 25*(12), 2301–2316.

Sun, D., & Pinker, R. T. 2003. Estimation of land surface temperature from a Geostationary Operational Environmental Satellite (GOES-8). *Journal of Geophysical Research, 108*(D11), 1–18.

Sun, D., Pinker, R. T., & Basara, J. B. 2004. Land surface temperature estimation from the next generation of Geostationary Operational Environmental Satellites: GOES M–Q. *Journal of Applied Meteorology, 43*(2), 363–372.

TET. 2019. *Emissivity coefficients materials.* Accessed on May 09, 2019. Retrieved from https://www.engineeringtoolbox.com/

Tomlinson, C. J., Chapman, L., Thornes, J. E., & Baker, C. 2011. Remote sensing land surface temperature for meteorology and climatology: A review. *Meteorological Applications, 18*(3), 296–306.

Tsou, J., Zhuang, J., Li, Y., & Zhang, Y. 2017. Urban heat island assessment using the Landsat 8 data: A case study in Shenzhen and Hong Kong. *Urban Science, 1*(10), 1–22.

UN DESA. 2018. *2018 revision of world urbanization prospects.* Accessed on September 27, 2018. Retrieved from https://www.un.org/development/desa/publicati ons/2018-revision-ofworld-urbanization-prospects.html

Valor, E., & Caselles, V. 1996. Mapping land surface emissivity from NDVI: Application to European, African, and South American areas. *Remote Sensing of Environment, 57*(3), 167–184.

Vogt, J. V. 1996. Land surface temperature retrieval from NOAA AVHRR data. In D'Souza, G., Belward, G., Alan, S., & Malingreau, J. (Eds.), *Advances in the use of NOAA AVHRR data for land applications* (pp. 125–151). Dordrecht: Springer.

Wan, Z. 2008. New refinements and validation of the MODIS land-surface temperature/emissivity products. *Remote Sensing of Environment, 112*(1), 59–74.

Wan, Z., Zhang, Y., Zhang, Q., & Li, Z. L. 2004. Quality assessment and validation of the MODIS global land surface temperature. *International Journal of Remote Sensing, 25*(1), 261–274.

Wang, J., Cui, Y., He, X., Zhang, J., & Yan, S. 2015. Surface albedo variation and its influencing factors over Dongkemadi glacier, central Tibetan Plateau. *Advances in Meteorology, 2015*, 1–10.

Wark, D. Q., Yamamoto, G., & Lienesch, J. H. 1962. Methods of estimating infrared flux and surface temperature from meteorological satellites. *Journal of the Atmospheric Sciences, 19*(5), 369–384.

Weng, Q., Rajasekar, U., & Hu, X. 2011. Modeling urban heat islands and their relationship with impervious surface and vegetation abundance by using ASTER images. *IEEE Transactions on Geoscience and Remote Sensing, 49*(10), 4080–4089.

Xiao, H., & Weng, Q. 2007. The impact of land use and land cover changes on land surface temperature in a karst area of China. *Journal of Environmental Management, 85*(1), 245–257.

Yamaguchi, Y., Kahle, A. B., Tsu, H., Kawakami, T., & Pniel, M. 1998. Overview of advanced spaceborne thermal emission and reflection radiometer (ASTER). *IEEE Transactions on Geoscience and Remote Sensing, 36*(4), 1062–1071.

Yanev, I., & Filchev, L. 2016. A comparative analysis between MODIS LST level-3 product and in-situ temperature data for estimation of urban heat island of Sofia. *Aerospace Research in Bulgaria, 28*, 77–92.

Yu, X., Guo, X., & Wu, Z. 2014. Land surface temperature retrieval from Landsat 8 TIRS comparison between radiative transfer equation-based method, split window algorithm and single channel method. *Remote Sensing, 6*(10), 9829–9852.

Zhou, J., Zhang, X., Zhan, W., & Zhang, H. 2014. Land surface temperature retrieval from MODIS data by integrating regression models and the genetic algorithm in an arid region. *Remote Sensing, 6*(6), 5344–5367.

7

Sustainable Development Goals (SDGs)

Disaster Mitigation in Flood-Prone Regions of India

CONTENTS

7.1 Introduction 127
7.2 Changing Climate Scenario 128
7.3 Sustainable Development Goals 2030 129
7.4 India and the Sustainable Development Goals 2030 130
    7.4.1 Urban Flood Vulnerability Zoning: A Case Study of Delhi 134
    7.4.2 Accessing Adaptive Capacity of a Region against Flood
        Vulnerability: A Case Study of Andhra Pradesh Capital
        Region 139
7.5 Conclusions 144
References 146

7.1 Introduction

Sustainability is the key concept towards the sustenance of natural resources by prudently managing land, water and environment through an integrated approach involving social, environmental and economic aspects. In this regard, the state-of-the-art endeavour is about global partnership and coordination to resolve matters for peoples' equity, peace and prosperity (UNSDSN, 2015). The United Nations Conference on the Human Environment in 1972 was the first international effort for such sensitive thinking followed by the emergence of the concept "sustainable development". Recently, the Sustainable Development Goals (SDGs) came with a holistic vision of development through building a robust system that can address environmental adversities by 2030. These international initiatives attempt to account for the socio-economic essentialities of the marginal population along with their environmental needs (Griggs et al., 2015; Tomislav, 2018).

The well-being and socio-economic profile of the community depend on ensuring preservation of natural resources. This dependency is complex and has been fundamental for subsistence living on earth. Sustainable development entails adapting progressive behaviour to ensure socio-ecological balance (Basiago, 1998).

The UN Millennium Development Goals were aimed to address 21 such key targets keeping in mind 63 critical indicators for various accomplishments. The success of this program paved way for Sustainable Development Goals 2030 (Kumar et al., 2016).

7.2 Changing Climate Scenario

The 21st century has witnessed large-scale urbanisation and industrialisation with globalisation The time period also marked the beginning of rapidly changing climate scenarios along with rising intensity of environmental turbulences, due to the continued unplanned developmental path. The world urban population rose from 39% in 1980 to 52% in 2011 with developing countries being the primary contributors in the urban hike (Chen et al., 2014). However, unlike developed countries, the rate of industrial growth within developing countries was unable to match the rate of rapid urban shift. As a result, the upcoming urban conurbations remained deprived of adequate urban infrastructure and utility services, and were predominated by urban poverty and climatic hazards (Chauvin et al., 2017).

Further, the global world limits the chances of budding economies to explore their traditional capabilities/skill sets. The countries remain constrained due to poor management of scarce natural resources coupled with the extensively growing demands of urban conurbations. Thus, the thriving need for immediate economic choices subdues the mindfulness for environmental-central development within spaces.

Nations within the tropical and subtropical regions are emerging as bigger victims of aggravated climate change phenomena – natural hazards, rising temperatures, soaring population and ecological degradation. As per the World Bank recent report, the tropics comprise of large range of developing and underdeveloped countries with some having a per capita GDP share of less than 8% (India) when compared to developed counterparts (United States) (World Bank, 2018). Therefore, vulnerabilities here get exceedingly complex due to the coupled effect of exposure and system incompetence. These countries have been undergoing a substantial socio-economic loss due to lack of awareness, technological know-how, greater densities at stake and inadequate system capability to adapt for the same (Beg et al., 2002; Kok et al., 2008).

For instance, the World Risk Index identifies disaster vulnerabilities through fundamental hazard–effect relationships between three major components: susceptibility of a region, its coping capability and adaptive capacity against the disaster causing stimulus (IFHV, 2018). The world risk and exposure map in the World Risk Report 2018 validates the grave effects of climate changes in regions facing a resource crunch with heavy population loads. Consequently, these urban centres have failed in fulfilling the infrastructure demands and needs for basic amenities. Therefore, the standalone targets of reducing poverty and hunger, enhancing liveability, and economic growth have been proposed with the target of adaptive capacity and resilient growth in the new global agenda of 2030.

7.3 Sustainable Development Goals 2030

The Sustainable Development Goals 2030 is an attempt for strengthening developing countries by considering revenue opportunities through non-polluting agents and environment-friendly sources. It talks about peace, harmony and equity among both developing and developed nations for resilient growth in this dynamic era of climate change, prioritising their social, economic, cultural and life-supporting needs (Griggs et al., 2015). The SDGs are in tune with the objectives of existing MDGs for disaster reduction.

Figure 7.1 outlines the adaptive capacity of a system towards vulnerability reduction. Sustainable goals address interlinked, multiple postulates, as described in Table 7.1.

The developing countries often function on the lines of rescue, relief and rehabilitation (3Rs), which help in effective management of post-hazard situations through emergency stabilising operations. However, the recent framework of disaster management extensively substantiates the role of disaster preparedness and an early-warning system in preventing such mishaps and reducing their overall tragic effect on life and property. For instance, the early warning system developed through Federal Emergency Management Agency (FEMA) guidelines in the United States forecasts an impending cyclone scenario followed by quick dissipation of information for evacuation of vulnerable areas.

The newly found agenda of Sustainable Development Goals 2030 comes with a broader purview and diverse framework of resilience that suits all levels of developmental criteria (as mentioned in Table 7.1). The perception of the goals have metamorphosed traditional disaster mitigation strategies to match well with the objective of resource mobilisation for adaptive growth within a system. The case examples from India will make it clear that policy-oriented changes based on innovative research outcomes have given an impetus to the global awakening. The "Economic Losses, Poverty & Disasters" report (1998–2017) identifies flooding-related disasters to be the most frequent occurring climate-related natural disasters in India and around the world, with frequency of about 43.4% during the past two decades affecting the lives of more than 2 billion (UNISDR, 2017).

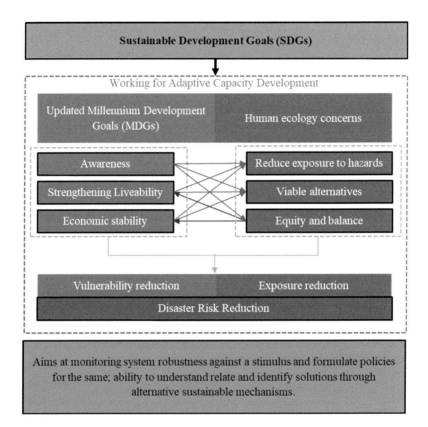

FIGURE 7.1
Components of SDGs.

7.4 India and the Sustainable Development Goals 2030

India opened its markets as part of the liberal economic reform subsequent to globalisation in 1990. The economic opportunities and socio-economic upgradation led to the diversification of secondary and tertiary services promoting rapid shift of employment centres from agro-rural to techno-urban.

The large-scale migration to the cities has enhanced basic demands for food and shelter, infrastructure needs and environmental status. The unorganised growth within urban centres devoid of strategic regulations resulted in modifications of aboriginal landscape forms, land uses typology and land cover of the region. One example is the blockage in the course of waterbodies due to construction on river beds and in catchment areas. Consequently, these have been leading to several incidents of urban floods in the country due to the lack of planning regulation enforcement and repeated incidents of anomalous weather events attributed to climate change. Cities including Ahmedabad (2001, 2017), Bengaluru (2005, 2009, 2013), Chennai (2004, 2005,

TABLE 7.1

SDG: Role in DRR through Adaptive Capacity Development

SDGs and Their Holistic Approach toward Adaptive Capacity
Development of System

There is not any substitute for
basic needs. The philosophical
concerns can only come into
practice if the basic necessities
of development are met.

Good health and social
well-being can ensure easy
escapes during a *hazards
moment* with smart
coordination's among
communities for symbiotic
living in limited availabilities.

It gives out the scope for
awareness about the *disaster
type* along with the respective
prevention and mitigation
mechanisms. It is also crucial to
understand the early signs of
hazard phenomena.

Women and children rights'
protection is a milestone
component for *adaptive capacity
development* and disaster risk
reduction.

It is linked to liveability that
again is an important factor in
restoring stability and
coordination among urban
units, thus developing their
capability to adapt.

Priority action in the present
climate change epoch. Investing
in renewable sources will lead
to higher *resilience* and reduce
hazard exposure.

(Continued)

TABLE 7.1 (CONTINUED)

SDG: Role in DRR through Adaptive Capacity
Development

This will ensure affordability to access municipal services and liveability standards to have adequate awareness and capability to sustain in *extreme situations*.

Technological innovation in robust, redundant and disaster-resilient infrastructure prevents loss of life and property at advent of *calamity*.

Sustainable and organised planning of cities will make them robust against *disasters* and promote easy rescues and relief for the same.

This is an amalgamation of all prior concerns regarding *disaster thinking* and its importance in human subsistence.

Protection of ecosystem services through introspective planning and land management is an immediate need to ensure robust territorial stability in this dynamic era of climate change. It shields one and all from the rising magnitude of *disaster events*.

Inter-, as well as intra-state coordination, is very important to meet on unified of resilience and redundant city-systems against the *disaster-causing impacts of climate change*.

2015), Delhi (2003, 2009, 2010, 2013, 2016), Gandhinagar (2017), Guwahati (2010, 2011, 2015, 2016, 2017), Hyderabad (2000, 2001, 2002, 2006, 2008), Jamshedpur (2008), Kolkata (2007), Mumbai (2005, 2007, 2015, 2017), Srinagar (1992, 2014, 2015) and Surat (2006) have been victims of urban flooding in the recent past (Sarmah & Das, 2018; Ramachandra, et al., 2012).

The 73rd constitutional amendment in 1992 realised the significance of stakeholders' participation in resource management, policy reform formulation and implementation through the decentralised administrative mechanisms. The rising challenges in cities due to haphazard urban growth, an escalated greenhouse gas footprint and consequent changes in the climate requires precise understanding of causes and local alternatives for better growth (Mahadevia, 2001).

The National Institution for Transforming India (NITI Aayog) has been shouldering the responsibility of implementing SDGs in the policy framework. The mission of every interlinking project re-establishes the need for "adaptive capacity development" as the issue of prime importance in the road to sustainability. The changes in the climate have irreparable effects on the sustenance of natural resources. It has limited the opportunity of improvisation beyond human control. The Combat Climate Change Action Plan goal 13 has resulted in various environment improvement missions such as the National Action Plan on Climate Change and National Mission for a Green India that aim at resilient growth with adequate capability to withstand climate-related hazards in all regions of the country, at different tiers of governance (Sharma & Tomar, 2010).

The postulates such as the National Solar Mission and National Mission for Enhanced Energy Efficiency connects to the action plan of goal 7 for sustainable development. It is based on addressing the energy needs of countries through exploring technological advancement in renewable energy sectors. In addition, the goal aims to promote greenhouse gas (GHG) reduction aiding other simultaneous columns of development. The National Mission on Sustainable Habitat, National Water Mission, National Mission for Sustaining the Himalayan Ecosystem and National Mission on Strategic Knowledge for Climate Change cumulatively aim for adaptive capacity development and effective climate change-related planning within the country, focusing on women, youth, and local and marginalised communities.

The agenda of resolving climate-related vulnerabilities due to flood-related disasters with the help of a sustainable planning framework are meticulously explained in the following two case studies. Both examples are at distinct scales and take diverse routes towards the common perspective of community resilience, adaptive growth and people-centred development (Mahadevia, 2001; Sathaye et al., 2006). The first case study centres around city-level problems due to urban floods. The second is a regional-level assessment of the adaptability required for resilient growth in coming years. Vulnerability can be referred to as sensitivity to hazards causing stimulus as a result of exposure and defencelessness (Birkmann, 2006). The case studies deal with flood-risk preparedness by understanding the spatial distribution of vulnerability within a region. In lines with the Hyogo and Sendai framework of disaster-risk resilience, they exhibit the use of geospatial technology in determining risk profiles for disaster-prone spaces.

7.4.1 Urban Flood Vulnerability Zoning: A Case Study of Delhi

India has been seeing massive urban growth for the past two decades. The growth character has however, been skewed and devoid of "development". The definition of urbanisation as per the population census of India is limited to attributes such as change in population density and shift in the occupational structure with no heed towards infrastructure needs and basic amenities of the growing population or the quality of life that an urban centre is supposed to offer (Singh & Rahman, 2018). Globalisation and consequent relaxations in the Indian market in the 1990s witnessed an unprecedented rural influx to urban pockets hoping for better living opportunities. This resulted in the growth of peri-urban pockets with a discrete settlement pattern. Loss of ecology and disregarding hydrologic regimes in the rapid urbanisation process led to expansive encroachment of ecologically sensitive areas leading to severe repercussion such as urban floods (Table 7.2).

Urban floods are situations caused due to inundation of basic infrastructure services in a built environment. City life comes to a halt, and this is usually the result of an incapacitated urban drainage network that fails to accommodate excess surface runoff during a high-intensity and short-duration rainfall event (Wheater, 2006). As mentioned earlier the primary reason for such flooding is a failure of the catchment to retard water velocity coupled with the blockage of storm-water channels and disappearance of lakes. However, flooding of small streams within a settlement zone, construction within the river catchment area, alterations in cloud formation and rainfall, coastal flooding and high-velocity tidal events are also responsible for the flooding of urban growth centres (Freitag et al., 2018). Urban flooding is a cumulative result of both natural and anthropogenic activities. For instance, changing climate scenarios are likely to alter the rainfall intensity over a given period and hence the peak discharge and runoff volume may exceed the capacity of conventionally designed drainage networks.

Similar to this, rapid urban development has transformed the natural land use arrangements. The encroachment of catchment areas and rising imperviousness due to change in the land use gives a limited chance of water infiltration causing high rates of runoff. The natural drainage pattern of the cities has also been blocked or narrowed and concretised due to lack of scientific planning and many times the low-lying catchment area and waterbodies become a dumping ground to solid waste and other municipal waste from the city itself.

The unplanned urban growth within the cities has consumed the benefits of urban living (Mohan et al., 2011). Urban floods pose a direct threat to

TABLE 7.2

Major Cities of India Facing Urban Floods and Their Causes

Name of the State	Affected Area	Years of Flooding (Urban Floods)	State's Level of Urbanisation in per cent		Causes
			2001	2011	
Jammu and Kashmir	Jammu and Srinagar	2014	24.81	27.21	The collapse of natural drainage, unplanned urbanisation, disappearing waterbodies, conversion of wetlands in dumping site
Delhi NCR	Delhi	2013	93.18	97.50	Reducing wetlands, unplanned urban development
Rajasthan	Jaipur, Udaipur		23.39	24.89	Pollution and encroachment on lakes, silting of waterbodies and weak embankments
Gujrat	Ahmedabad, Surat, Vadodara	2000	37.36	42.58	Encroachment of lakes, lack of climate-sensitive planning
Assam	Guwahati		12.9	14.08	Depleting wetlands due to encroachment and illegal construction activity
West Bengal	Kolkata		27.97	31.89	The disrepair of drainage system, silting and blocking of drains, poor solid waste management
Maharashtra	Mumbai, Kolhapur		42.43	45.23	Encroachment upon water body and wetland, clogged drains
Chhattisgarh	Raipur		20.09	23.24	19.03% of wetland area was lost between the year 1976 and 2006
Karnataka	Bengaluru		33.99	38.57	Unplanned urban growth; encroachment on waterbodies; loss of interconnectivity among waterbodies; narrowing and concretising natural drains; removal of 79% waterbodies during the past five decades; sustained dumping in lake catchment areas, drains and in the lake bed

(Continued)

TABLE 7.2 (CONTINUED)

Major Cities of India Facing Urban Floods and Their Causes

Name of the State	Affected Area	Years of Flooding (Urban Floods)	State's Level of Urbanisation in per cent		Causes
			2001	**2011**	
Andhra Pradesh and Telangana	Hyderabad	2000,	27.3	33.4	Unplanned urban development; city lost 404 lakes between 1982 and 2012; in the last three decades, the percentage area under waterbodies has reduced by 5
Tamil Nadu	Chennai	2015	44.04	48.45	Reduction in the area of marshland, waste disposal and construction on wetland area, lack of eco-sensitive planning in the city

Source: Down to Earth, 2016.

human life and well-being of the people. It damages the existing infrastructure such as roads, bridges, buildings, railway lines, etc., and corresponds to the related physical discomforts and economic losses. Thus, they pose an incomprehensible challenge for city governance.

Urban flood vulnerability zoning is a non-structural, flood-risk mitigation approach to categorically earmark flooding scenarios within an urban region. The zoning areas do exhibit the propensity of being flooded at time of influx. The spatial analysis is done for evaluating conflicting points within the system that have been adding to its sensitivity for disaster, causing stimulus. Temporal remote sensing data provide ease in assimilating changes within the urban morphology, which aid in sensible policy formulation. The scaled results provide details for introspection and intervention that can help in reducing vulnerability or system failures.

Delhi, the national capital of India has a population of about 22.2 million residents within the National Capital Region (NCR) that extends up to 1484 sq km (Swerts et al., 2014). The prime capital city of India has witnessed an expansive growth of about 4.1% that is highest among the Indian metropolis. It is the eighth-most populous city in the world and the second most in India. Delhi is located on the right bank of the river Yamuna at the periphery of Gangetic plains.

The glorious city and innumerable cultural heritages have played an administrative centre for several ruling dynasties in the past. Delhi was the nodal platform for trade and commerce across the country and has experienced a sudden influx in population after independence. The city population doubled after the partition of India, forcing it to accommodate the same

TABLE 7.3

Area, Population and Population Density of Delhi

Area (sq km)	Year	Population (lakhs)	Population Density (persons/sq km)
1484	1971	40,65,698	1207
	1981	62,20,406	1899
	1991	94,20,644	2804
	2001	138,50,507	4371
	2011	167,53,235	5726

Source: Census India – 1971, 1981, 1991, 2001 and 2011.

in a sprawled, expansive and unplanned growth. However, apart from the post-independence resettlements, Delhi has been attracting migrants from the nearby adjoining states of Bihar, Uttar Pradesh, West Bengal, Haryana, Punjab and Rajasthan (Mohan et al., 2011) (Table 7.3).

Due to uncontrolled and unregulated settlement patterns, the national capital has been facing an increased vulnerability to urban flooding (De et al., 2013; Sarmah & Das, 2018). The flood vulnerability zoning for Delhi includes three basic steps. The first step is identification of the actual vulnerability profile by delineating the flood areas with the help of primary and secondary data sources. This included the creation of influencing thematic layers for the same. The second step was based on a spatial reclassification of these maps on a vulnerability score of 1–10. The final step integrated these layers into a zonal vulnerability profile map using the multi-criteria evaluation (MCE) method and weighted overlay technique (Tables 7.4 and 7.5).

The temporal vegetation cover analyses, shown in Figures 7.2 and 7.3, with the help of the normalised difference vegetation index (NDVI) and temporal land uses in Delhi (with the NCR) evidently prove the trends of growing urbanisation with the depleting vegetation cover. The percentage area under vegetation has significantly decreased from 30.16% (in 2003) to 20.03% (in 2017). This can be described as the consequence of a growing urban footprint in the region and related anthropogenic developments. The areas under waterbodies also show a declining trend from 1.24% to 1.01%. This can be correlated to the encroachments upon the floodplains of the Yamuna River. The rapid transformation of an area into paved urban surfaces and

TABLE 7.4

NDVI Values Reflecting the Areas under Vegetation and Non-Vegetation

Land Cover (%)	2003	2010	2017
Vegetation	37.26	33.87	28.74
Non-Vegetation	62.74	66.13	71.26

TABLE 7.5

Land Use Statistics of the Classified
Images

Land Use (%)	2003	2010	2017
Urban	21.26	26.83	32.37
Vegetation	30.16	28.16	22.03
Water	1.24	1.17	1.01
Others	46.97	43.84	44.59

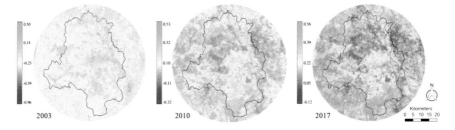

FIGURE 7.2
Land cover for time period 2003–2017.

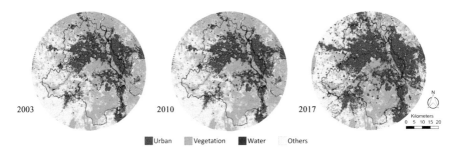

FIGURE 7.3
Land uses (with four different categories) for the time period 2003–2017.

built forms signifies thriving feasibility for urban flooding within the capital
region.

The influencing thematic maps are comprised of digital elevation model
(DEM) for the region, the slope of the area, and the natural drainage flow
along with the blocked locations. The layers are also comprised of administrative ward jurisdictions; waterlogging locations; rail and road networks;
waterbodies in the region; major stations; social amenities; green areas; and
locations of industries, political and religious places. Each of these spatial
layers was transformed into a "distance map" using the interpolation technique to identify influential curves. The influence of the factor is said to nega-
tively correlate to its rising distance. This step was followed by MCE analysis
of the distance map into a "suitability map" using the Boolean approach.

The binary profile considered the included area as 1 (red colour) and the excluded area as 0 (black colour). For instance, all slopes greater than 2% have been considered significant in elevating urban flood phenomena. The suitability map was generated using a fuzzy approach so that all the layers involved in MCE analysis can be brought down or normalised to a single equivalent scale ranging from 0 to 255. The 0 here is the area of least influence from a factor vice versa at 255.

Table 7.6 shows the global weights of all considered factors derived on the basis of pairwise comparison through an analytical hierarchical process with a consistency ratio of 0.06. The global weights were used for valuation in the multi-criteria decision analysis (MCDA) process to integrate the layers as the final "urban flood vulnerability zoning map". The map categorised urban flood vulnerability in the NCR as low, moderate, high and very high zones, susceptible to urban flood incidents (Figure 7.4).

7.4.2 Accessing Adaptive Capacity of a Region against Flood Vulnerability: A Case Study of Andhra Pradesh Capital Region

The objective of disaster management has been multi-faceted and is incomplete without integrating stakeholders' vision for safety in an extreme situation. It covers the broad framework of pre- and post-disaster mitigation agendas based on dynamics of the recurring situation. Therefore, data acquisition and processing of the same as important information code is of the utmost priority for ensuring resilience of people and surrounding. The information code can be the baseline scenario for the region that plays a significant role in closing the gaps in the process of resilience development and an important subsidiary for projected visuals.

TABLE 7.6

Weights of Influencing Factors

Influencing Factor	Weight
Drain block sites	0.1883
Social amenities	0.0371
Industries	0.0807
Green areas	0.0164
Political places	0.0807
Religious places	0.0371
Rail and road network	0.1499
Major stations	0.0371
Waterbodies	0.0473
Waterlogged areas	0.2146
Low population density	0.0141
Medium population density	0.0289
High population density	0.0678

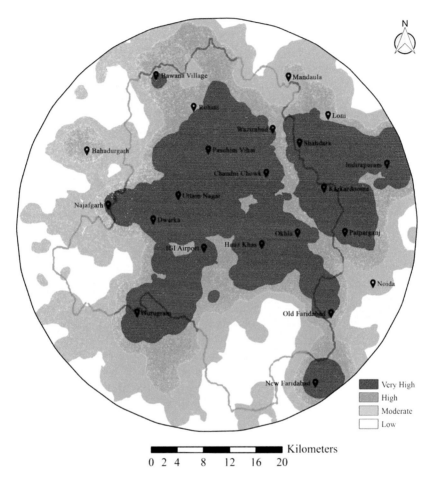

FIGURE 7.4
Urban flood vulnerability map for the case area.

Adaptive capacity assessment is one such mitigation tool depending on baseline data integration process to analyse the coping capability of a community against a disaster stimulus (Batica & Gourbesville, 2014). The disaster risk refers to the intensity of a hazard, and robustness of the system to withstand and rejuvenate afterwards. The case study tries to report the system conditions during vulnerable times.

The Andhra Pradesh Capital Region (APCR) is an area of 8603.32 sq km and is a newly delineated greenfield project of Andhra Pradesh state government for developing the new state capital "Amaravati". The capital region boundary comprises a total of 56 revenue divisions called "mandals" from the adjoining Krishna and Guntur districts of Andhra Pradesh. Amravati, with an area of 217 sq km, is at a close proximity to the city of Vijayawada and is located on the bank of the Krishna River that bifurcates the entire capital region in two districts, namely Krishna and Guntur (Andhra Pradesh Capital Region Development Authority, 2017) (Figure 7.5).

APCR is located near the Krishna River delta at the eastern coast of India. The closer proximity to the sea makes the downstream area highly susceptible to flood-related disasters. There have been several incidents of flooding in the past decade, mostly due to the occurrence of severe rainfall in the upstream. The region is attributed with a chain of non-perennial rivers, overlaid with several multi-purpose dam projects. Rising intensity of the rainfall in the past few years (in tune with the pattern of climate change) has caused uncertainty in the estimation of runoff. In October 2009, the river basin received 25.4 lakh cusec of water, much higher than the safety levels of past flood discharge record of 9.32 lakh cusec in the year 1988 (APWRDC, 2009). The floodgates of Nagarjuna Sagar were opened to release excess water, which resulted in flooding downstream. An agrarian economy dominates the region, and it is profoundly affected due to frequent flood events leaving the poor farmers in distress. The census also reports suffering among the marginalised communities due to lack of equitable distribution of public infrastructure and limited municipal services.

The impoverished population lacks affordable safe shelter, good food, drinking water, a medical facility and any kind of awareness regarding preparedness for adverse situations. Analogous to the objective, the case study assembles socio-economic as well as spatial data indices to assess the risk that the study area may be vulnerable to. Mandal-wise evaluation was done with the simulation of future scenarios considering a 2050 land-use prediction map.

Land use analyses involved a temporal land use classification of Landsat satellite data for the years 2011 and 2018, downloaded from United States Geological Survey (USGS) public portal. The Gaussian maximum likelihood method (GLMC) was used for the supervised interpretation of data sets into five prominent classes: urban, forest, cropland, water and other land uses. These preliminary maps were used as input for the cellular automata "non-agent"–based model to project land use by 2050 (Figure 7.6 and Table 7.7).

Census data of year 2011 was considered for socio-economic indices, including (1) accessibility, (2) population density, (3) female education, (4) health infrastructure, (5) number of kutcha houses in a mandal, (6) number of marginal workers in a mandal, and (7) average per capita income within a mandal. The indicators were based on the series of interlinking factors mentioned within the Andhra Pradesh state government document "State Action Plan on Climate Change". The spatial configurations for parameters such as (8) elevation of the region and (9) stream distances were marked with help of the ASTER Global Elevation Model V002 (30 m) data sets. Landsat series satellite data of TM and ETM sensors were used to obtain land use information for the years 2011 and 2018.

The assessment was based on computing census values for each mandal followed by rational assignment of comparative scores for the same. The score

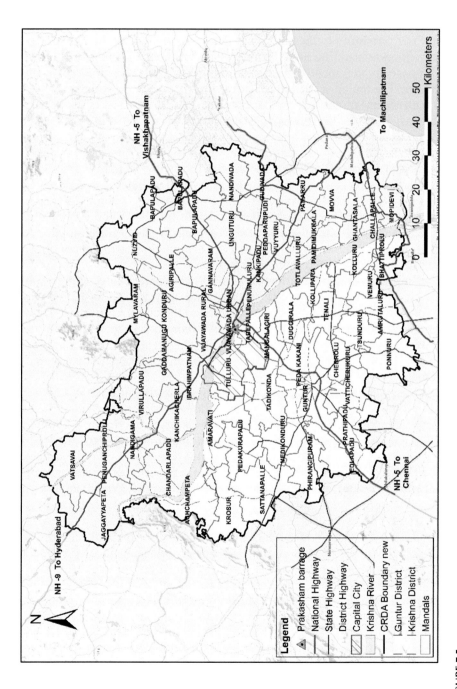

FIGURE 7.5
Study area – APCR.

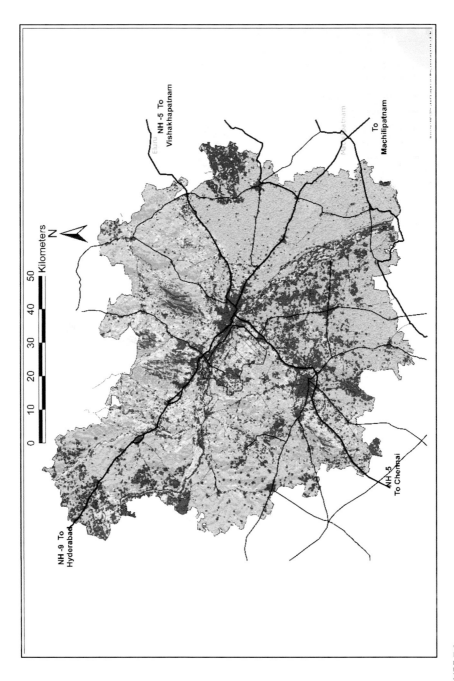

FIGURE 7.6
Proposed land use map of APCR 2050.

TABLE 7.7

Land Use Classification Details

Land Use Class	Land Uses Included in the Class
Urban	Residential area (rural as well as the urban settlements), industrial area, and all paved surfaces and mixed pixels having built-up area
Forest	Protected forest area in the region
Agriculture	Cropland (sown and unsown), nurseries
Waterbody	River, aquacultures, tanks, lakes, reservoirs
Others	Rocks, quarry pits, open ground at building sites, barren land, kutcha roads, river bed (dry)

distribution lies on a Likert scale of 1 to 5 depending upon the progressive or regressive fallout of spatial and non-spatial factors on the flood risk positions. For example, a rise in elevation is a positive attribute for resilience and negative for disaster risk. Similarly, the higher the number of kutcha houses, the higher the risk factors of the mandal. The score value of 1 is assigned for minimum resistance to hazards related stimulus, to 5 for a higher-order risk level. The priorities among the chosen risk assessment parameters were decided through pairwise comparison on a 'Saaty' 9-point multi-criteria and decision making (MCDM) scale using an analytical hierarchical process (AHP) (Saaty, 2008). The spatial vulnerability of the APCR was anticipated using a "weighted sum" regime. The result was assimilated with the projected land use of 2050 to clearly identify future growth centres highly susceptible to hazard risk in the business as usual (BAU) scenario (Figure 7.7 and Table 7.8).

7.5 Conclusions

The sustainable development approach ensures the sustenance of natural resources for a safer future (Swart et al., 2004; Dovi et al., 2009). Marginal economic profiles are due to poor resource management and limited financial growth. The Sustainable Development Goals for 2030 aims towards restoring developmental equity on the planet accounting for climate changes and associated disaster vulnerabilities. The goals are to register policy level integrity for a robust system that has a higher capability to withstand climate changes with the rising intensity of natural hazards. The cases studies provide insights into the existing scenarios within developing states that have started realising stresses due to changes in the climate pattern. This would help in evolving progressive mitigation protocols that help in building a stronger adaptive capability against flood-related hazards. The study acknowledges the potential of temporal remote sensing data in spatial planning initiatives to develop sustainable policies.

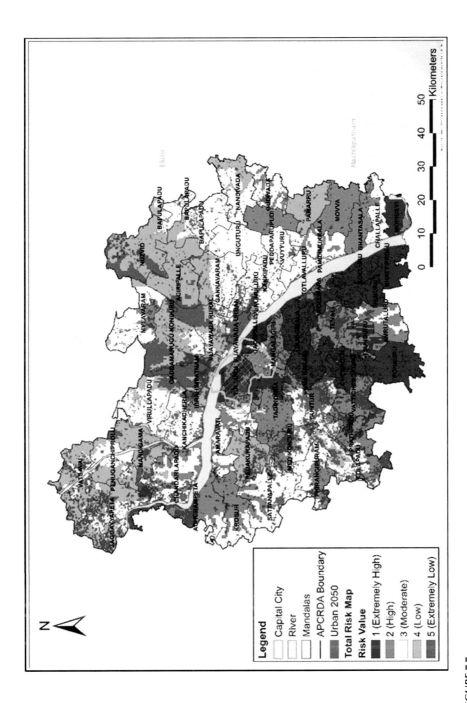

FIGURE 7.7
Total risk map of APCR.

TABLE 7.8

Evaluating Parameters with Their Global Weights

Parameter	Relevance to the Context	Effect on the Regional Vulnerability	AHP Score
Accessibility	Accessibility within a mandal determines resource mobility within it at the occurrence of extreme events. It can be related to both evacuation and relief processes.	Regressive	0.111
Population density	The higher the population per unit area, the more they are exposed to hazard threats.	Progressive	0.021
Female education	It is concerned with the awareness among the women and children about the risks associated to natural disasters. Female mortality during a disaster is 14 times higher than that of men. The Sendai Framework objectively puts forth this issue of gender inequality in the global motto of disaster risk reduction (UNDP, 2013).	Regressive	0.271
Health infrastructure	Presence of public-sponsored health infrastructure in the region ensures that the marginalised citizens can hope for a better recovery in a disaster-struck environment.	Regressive	0.146
Number of kutcha houses in a mandal	Kutcha houses are non-engineered houses such as thatched huts usually made of mud, straw, bamboo or other non-durable material that have a higher tendency to crumble during torrential rain and flooding (Census of India, 2011a; Census of India, 2011b).	Progressive	0.068
Number of marginal workers in a mandal	It is the worker population that works for a duration of less than 6 months in a year (GOI, 2011). An agrarian economy has landless labourers who do not have permanent employment or source of income most of the year.	Progressive	0.036
Income	The average per capita income in mandals is directly related to affordability of all other basic necessities essential for the survival of a person.	Regressive	0.047
Distance from the stream	The proximity to the river is associated to flood vulnerability in case of overflow, surge or erosion of edges during peak flows.	Regressive	0.035
Elevation	The settlements at higher elevations are less likely to be affected by floods than those living on floodplains or river deltas.	Regressive	0.261

References

Andhra Pradesh Capital Region Development Authority. 2017. *Facts & figures. Andhra Pradesh capital region development authority.* Accessed on November 24, 2017. Retrieved from https://crda.ap.gov.in/APCRDADOCS/

Andhra Pradesh Water Resource Development Corporation. 2009. *Managing historic flood in Krishna River basin - An experience of averting catastrophe* (Vol. 91). Andhra Pradesh Water Resource Development Corporation. Retrieved from https://www.indiawaterportal.org/

Basiago, A. D. 1998. Economic, social, and environmental sustainability in development theory and urban planning practice. *Environmentalist*, *19*(2), 145–161.

Batica, J., & Gourbesville, P. 2014. Flood resilience index-methodology and application. In *Proceedings of the 11th International Conference on Hydroinformatics*, New York.

Beg, N., Morlot, J. C., Davidson, O., Afrane-Okesse, Y., Tyani, L., et al. 2002. Linkages between climate change and sustainable development. *Climate Policy*, *2*(2–3), 129–144.

Birkmann, J. 2006. *Measuring vulnerability to natural hazards: Towards disaster resilient societies*. United Nations University Press.

Census of India. 2011a. *District census handbook Krishna village and town directory*. Accessed on June 10, 2019. Retrieved from http://censusindia.gov.in/2011census/dchb/2816_PART_A_DCHB_KRISHNA.pdf

Census of India. 2011b. *District census handbook Guntur village and town directory*. Accessed on June 12, 2019. Retrieved from http://censusindia.gov.in/2011census/dchb/2816_PART_A_DCHB_KRISHNA.pdf/

Chauvin, J. P., Glaeser, E., Ma, Y., & Tobio, K. 2017. What is different about urbanization in rich and poor countries? Cities in Brazil, China, India and the United States. *Journal of Urban Economics*, *98*, 17–49.

Chen, M., Zhang, H., Liu, W., & Zhang, W. 2014. The global pattern of urbanization and economic growth: Evidence from the last three decades. *PLoS One*, *9*(8), e103799.

De, U. S., Singh, G. P., & Rase, D. M. 2013. Urban flooding in recent decades in four mega cities of India. *Journal of the Indian Geophysical Union*, *17*(2), 153–165.

Dovì, V. G., Friedler, F., Huisingh, D., & Klemeš, J. J. 2009. Cleaner energy for sustainable future. *Journal of Cleaner Production*, *17*(10), 889–895.

Down to Earth. 2016. *Why urban India floods: Indian cities grow at the cost of their wetlands*. Accessed on April 24, 2019. Retrieved from https://cdn.downtoearth.org.in/uploads/0.56859500_1457688308_Preview.pdf

Freitag, B. M., Nair, U. S., & Niyogi, D. 2018. Urban modification of convection and rainfall in complex terrain. *Geophysical Research Letters*, *45*(5), 2507–2515.

Griggs, D., Stafford-Smith, M., Gaffney, O., Rockstrom, J., Ohman, M. C., et al. 2015. Sustainable development goals for people and planet. *Theory and Science*, *495*, 305–307.

Institute for International Law of Peace and Armed Conflict (IFHV). 2018. *World risk report 2018*. Accessed on May 8, 2019. Retrieved from https://reliefweb.int/sites/reliefweb.int/

Kok, M., Metz, B., Verhagen, J., & Van Rooijen, S. 2008. Integrating development and climate policies: National and international benefits. *Climate Policy, 8*(2), 103–118.

Kumar, S., Kumar, N., & Vivekadhish, S. 2016. Millennium development goals (MDGS) to sustainable development goals (SDGS): Addressing unfinished agenda and strengthening sustainable development and partnership. *Indian Journal of Community Medicine: Official Publication of Indian Association of Preventive & Social Medicine*, *41*(1), 1.

Mahadevia, D. 2001. Sustainable urban development in India: An inclusive perspective. *Development in Practice, 11*(2–3), 242–259.

Mohan, M., Pathan, S. K., Narendrareddy, K., Kandya, A., & Pandey, S. 2011. Dynamics of urbanization and its impact on land-use/land-cover: A case study of megacity Delhi. *Journal of Environmental Protection, 2*(09), 1274.

Ramachandra, T. V., Aithal, B. H., & Kumar, U. 2012. Conservation of wetlands to mitigate urban floods. *Journal of Resources, Energy and Development, 9*(1), 1–22.

Saaty, T. L. 2008. Decision making with the analytic hierarchy process. *International Journal of Services Sciences, 1*(1), 83–98.

Sarmah, T., & Das, S. 2018. Urban flood mitigation planning for Guwahati: A case of Bharalu basin. *Journal of Environmental Management, 206*, 1155–1165.

Sathaye, J., Shukla, P. R., & Ravindranath, N. H. 2006. Climate change, sustainable development and India: Global and national concerns. *Current Science-Bangalore-, 90*(3), 314.

Sharma, D., & Tomar, S. 2010. Mainstreaming climate change adaptation in Indian cities. *Environment and Urbanization, 22*(2), 451–465.

Singh, C., & Rahman, A. 2018. Urbanising the rural: Reflections on India's National Rurban Mission. *Asia & the Pacific Policy Studies, 5*(2), 370–377.

Swart, R. J., Raskin, P., & Robinson, J. 2004. The problem of the future: Sustainability science and scenario analysis. *Global Environmental Change, 14*(2), 137–146.

Swerts, E., Pumain, D., & Denis, E. 2014. The future of India's urbanization. *Futures, 56*, 43–52.

Tomislav, K. 2018. The concept of sustainable development: From its beginning to the contemporary issues. *Zagreb International Review of Economics & Business, 21*(1), 47–74.

UNDP. 2013. *Human Development Report 2013*. Accessed on June 10, 2019. Retrieved from http://hdr.undp.org/sites/default/files/reports/14/hdr2013_en_complete.pdf

UNISDR. 2017. *Economic losses, poverty & disasters*. Accessed on June 21, 2019. Retrieved from https://www.unisdr.org/files/61119_credeconomiclosses.pdf

UNSDSN. 2015. Getting started with the Sustainable Development Goals. Accessed on September 10, 2019. Retrieved from https://sustainabledevelopment.un.org/index.php?page=view&type=400&nr=2217.

Wheater, H. S. 2006. Flood hazard and management: A UK perspective. *Philosophical Transactions of the Royal Society A: Mathematical, Physical and Engineering Sciences, 364*(1845), 2135–2145.

World Bank. GDP per capita, PPP (current international $). Accessed on August 15, 2019. Retrieved from https://data.worldbank.org/indicator/NY.GDP.PCAP.PP.CD

8

Spatial Decision Support System (SDSS) for Urban Planning

CONTENTS

8.1 Introduction..149
    8.1.1 Rule-Based Models 151
    8.1.2 Agent-Based Models (ABM)..151
8.2 Spatial Decision Support System (SDSS) 152
8.3 Bangalore (Garden City) to Bengaluru (Unliveable City) 154
    8.3.1 Evolution of Bengaluru: From a Hamlet to Humongous
        Metropolis..155
    8.3.2 Status of Lakes and Trees in Bengaluru 156
8.4 SDSS Framework for Urban Planning 158
    8.4.1 Scenario 1: Business as Usual Scenario......................161
    8.4.2 Scenario 2: Sustained Growth Scenario.....................162
8.5 Conclusions...168
Acknowledgements 169
References...169

8.1 Introduction

Urbanisation is an irreversible process where the landscape tends to transform from rural to urban with paved surfaces, etc. (Voigtlander et al., 2008). Globally, the urban population has increased from 30% (1950) to 54% (2014) and is predicted to increase to 66% by 2050 as per the United Nations (UN DESA, 2018). India has witnessed an increase in urban population from 17.2% (1951) to 31.15% (2011) (Census of India, 2011) and is expected to increase to 46% by 2030 (TERI, 2019). Globalisation and consequent market reforms in the 1990s witnessed a spurt in industrialisation (manufacturing, IT, etc.) leading to an influx of population from rural areas (villages) to urban areas (towns/cities), along with an escalation in demand for natural resources, infrastructure and basic amenities. Unplanned urbanisation has been increasing in urban areas, encroaching other landscape elements, and affecting the local environment and ecology. This makes it necessary to understand the drivers of change, rate of change and likely future trends by integrating spatial and attribute databases and simulation models with policies, plans, etc. through spatial decision support systems (SDSS).

Urban centres are often branded as economic growth engines (Cerreta & De Toro, 2012) and an influx of people from rural areas/villages to urban areas would increase the social, economic and environmental problems. Unplanned and unorganised urbanisation often leads to growth in peri-urban areas and regions are devoid of either appropriate infrastructure or basic amenities, as urban planners often fail to understand the dynamics of such growth (Bharath et al., 2018a). Opportunities for jobs and education; relatively better infrastructure; transportation convenience; government services; industries; institutions; and political, cultural and other socio-economic factors act as catalysts of urbanisation. Figure 8.1 describes the causes and effects related to urbanisation. The process involves conversion of various land uses to paved surfaces such as built-up areas (laterally or vertically) and roads to cater to the growing needs of the society. The regions undergoing unplanned rapid urbanisation are witnessing demographic changes hampering the sustenance of local natural resources, alteration in air quality (Grimm et al., 2008), local ecology, and the quantity and quality of water (surface and subsurface quality) (Vinay et al., 2013). This necessitates incorporation of sustainable land use management plans with the understanding of landscape dynamics and the role of agents.

The conventional approaches to understand landscape dynamics are through land-based surveys, which are time- and resource-consuming. The availability of multi-resolution (spatial, spectral and temporal) remote sensing data with advances in GIS (Geographic Information System) technologies has proved reliable and economical for understanding landscape dynamics (NASA, 2001; Burrough, 1986; Ramachandra & Bharath, 2012; Ramachandra et al., 2012; Chen et al., 2000; Ji et al., 2001).

Geovisualisation of landscape dynamics is being attempted through diverse statistical techniques such as regression techniques, fuzzy logic, neural networks, Bayesian, binary modules, rules, etc., which are embedded

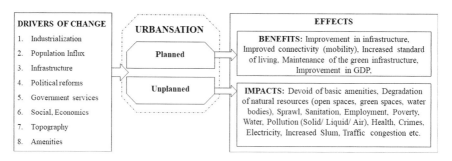

FIGURE 8.1
Urbanisation process and its implication.

with the spatial data through rule-based and agent-based models such as cellular automata, Markov chains, SLEUTH, SLUCE, Dynamica, Dyna-CLUES, ANN, multi-regression, analytical hierarchical process, multi-criteria evaluation, CAPRI-Spat and genetic algorithms (Arsanjani et al., 2013; Rafiee et al., 2009; Bharath et al., 2016; Siddayao et al., 2014; Jain et al., 2017; Grekousis et al., 2013; Mohammady et al., 2014; Taubenbock et al., 2009; Taubenbock et al., 2012; Sun et al., 2014; Mena et al., 2011; Bharath et al., 2014).

8.1.1 Rule-Based Models

Multiple rules used in order to simulate the future scenario through historical data sets has proved to be one of the best modelling techniques for urban growth simulation. Cellular automata (CA) is used to predict the state of the cell based on the previous state of the cells within a neighbourhood, using a set of transition rules. Markov chains are used to provide probabilistic transition events. Rule-based models use Markov chains integrated with cellular automata models. Although CA-Markov gives promising results, it fails to achieve accurate results due to non-accounting of agents – urban driving forces (Wang et al., 2013; He et al., 2013).

8.1.2 Agent-Based Models (ABM)

The area of influence is determined through interaction of the driving forces in the region. Land use changes are dependent on the influence of the drivers and neighbourhood interaction and the base modelling criteria. An agent-based model (ABM) uses factors causing or constricting growth of a landscape. The ABM technique generates the drivers of urban growth interaction through using fuzziness in the data sets and using the analytical hierarchical process (AHP) to account for its weight-based interactions through the CA-Markov process (Bharath et al., 2013). The ABM method is effective in simulating the changes based on agents. ABMs weigh/rank the growth factors and constraints as reflected by real-world scenarios to develop site suitability maps in order to model the land use. Site suitability maps provide the transitional areas (describing where the particular land use) with the probability to change or retain their state. The site suitability maps are combined with the CA-Markov in order to simulate and predict the land use dynamics. The modelling based on ABM has emerged as a promising approach for understanding the complex urban processes. Integration of fuzzy rules and AHP with ABM aid in defining the influence of the agents in urban simulation. The agent gradually transfers from a non-member to a member. This gradual transfer happens through a function called the membership function that is used to characterise various sets of agents that interact with a particular urban land use. These agents' interaction is treated as a fuzzy rule to define somewhat influential to least influential and a precise numerical of 8 bits to normalise the influence. This involves building a fuzzy network to extract fuzzy rules through an expert system. Zadeh (1965) introduced fuzzy sets to describe membership to a set in a partial order. The fuzzy sets of inputs through fuzzy rules are combined to represent the agent's strength, which is then combined with AHP to define the site suitability with interaction of the current land use to weigh characteristics of diverse opinions in a complex environment. This is a two-step process involving the assessment of effectiveness and alternatives. AHP has been a commonly used multi-criteria evaluation (MCE) technique for suitability of landscape through pairwise analysis. Integration of artificial intelligence with the MCE technique improves efficiency of modelling real-world situations and for understanding spatial patterns of urbanisation.

Land use dynamics information, simulation models, and prediction scenarios with spatial visualisation options are integrated in the spatial decision support system for visualising the current status, landscape dynamics (based on the past data) and likely implications of the decision towards formulating sustainable land use management strategies.

8.2 Spatial Decision Support System (SDSS)

An SDSS allows one to make choices or take decisions between two or more alternatives, by combining attribute data, spatial data (vector and raster), temporal data and decision logic as a tool (Crossland et al., 1995; Crossland, 2008; Sugumaran & DeGroote, 2010; Ramachandra et al., 2005). The process of decision making involves analysis of data, perspectives, methods, system dynamics, models, derived products, plans, scenarios, policy interventions, etc. (Ramachandra et al., 2017b).

An SDSS is "a coherent system of computer-based technology (hardware, software and supporting documentation) used by landscape decision makers as an aid to the regional decision-making in semi-structured tasks". The characteristics of SDSSs are (1) flexibility and adaptability to accommodate changes in the environment and the decision-making process of the user; (2) assist managers in their decision processes in semi-structured tasks/ problems for which formal models are useful, but where the planner's judgement is also essential; (3) support and enhance, rather than replace managerial judgement; (4) combine the use of models of analytical techniques with traditional data access and retrieval functions; (5) specifically focus on features which make them easy to use by non-computer people in an interactive mode; (6) organise data and models around decision(s) and are user-initiated and controlled; (7) present information in a flexible way to support the widely differing requirements and cognitive styles of users with the scope for interoperability of data formats; (8) use hierarchical design and are designed as model generators. That is, the model itself is built rather than the model building tools (Ramachandra et al., 2005; Ramachandra et al., 2017c).

The focus of an SDSS is on providing flexible tools for policy analysis and not on providing models to give answers to structured problems (Ramachandra et al., 2005). Indeed, the modelling tools are only one component of an SDSS, which may be described as comprising of three component subsystems (Ramachandra, 2009; Ramachandra et al., 2004):

1. Database management subsystem – Manages an integrated spatial and non-spatial database to drive all the models. Its purpose is to extract and combine information from a variety of sources, display the data structure to the user in a logical way, and handle personal and unofficial data, so that the user can experiment with alternatives based on personal judgement.

2. Model management subsystem – The DSS offers the user a number of modelling tools. The capabilities of this subsystem include creating new models, cataloguing and editing existing models, interrelating models by links through the database, and integrating small model "building blocks" into larger model systems.

3. Dialogue management subsystem (the user-system interface) – Includes a consistent and familiar interface, help functions, natural language messages (both in normal operation and error conditions), data entry forms, data entry validation and housekeeping functions (data back-ups, copying, deleting files, etc.). The SDSS design is guided by various methodological considerations, such as the scenario approach (simulates alternative land use scenario under a range of different assumptions).

A general framework of an SDSS is as depicted in Figure 8.2. The user, i.e., decision maker or analyst, can use the user interface (web server or offline/computer software) to interact with the various components of that support the SDSS. These components include the databases, knowledge bases, GIS tools, modelling techniques and scenarios. Databases store the multi-resolution spatial (raster/vector) data with the attribute information such as administrative boundaries and catchments. The knowledge base consists of information regarding various land use policies, city development plan (CDP), etc. GIS tools are used to interact with the spatial databases, analyse spatial data and the knowledge base, and create simulation models and scenarios to derive and visualise information. In the absence of the data/knowledge, the user interface works with the web servers to use the external data such as Web Map Service (WMS) and Web Feature Service (WFS) to fetch the information/derived products which would enable the user to make appropriate decisions.

In the current chapter, GIS and SDSS are used to understand the regional urban dynamics in Bengaluru (BBMP), and understand future growth using

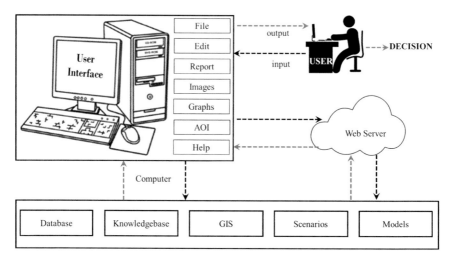

FIGURE 8.2
SDSS framework.

spatial (rule based, agent based) and non-spatial (regression) models using various scenarios. This understanding would help in the prudent management of natural resources (land) in Bengaluru through appropriate polices and framework.

8.3 Bangalore (Garden City) to Bengaluru (Unliveable City)

Bengaluru is the state capital of Karnataka State and the IT capital of India. Located in the Deccan Plateau of peninsular India, Bengaluru with the spatial extent of 741 sq km extends between 12.83°N, 77.45°E and 13.14°N, 77.78°E. Bengaluru was known as the Garden City of India for its lush green gardens, recreation spaces, salubrious climate, etc. The IT sector boom in the city laid foundation for city expansion through Special Economic Zones (SEZs), etc. which instigated the Garden City to Silicon City (Ramachandra et al., 2012; Bharath et al., 2013).

Bengaluru was ruled by various dynasties including the Chalukyas, Gangas, Hoysalas, Vijayanagara Empire to Kingdom of Mysore, followed by British rule until India's independence (Ramachandra et al., 2012). Later, the government of Karnataka formed the City Improvement Trust Board (CITB), which was responsible to plan and manage the city's infrastructure, growth, water supply, etc. CITB gave way to para-state agencies – BDA (Bangalore Development Agency) and BBMP (Bruhat Bengaluru Mahanagara Palike) subsequently (Figure 8.3). After India's independence, Bengaluru has seen large-scale development in terms of infrastructure,

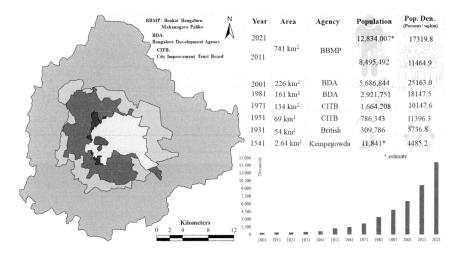

Year	Area	Agency	Population	Pop. Den. (Persons/sq.km)
2021			12,834,007*	17319.8
2011	741 km²	BBMP	8,495,492	11464.9
2001	226 km²	BDA	5,686,844	25163.0
1981	161 km²	BDA	2,921,751	18147.5
1971	134 km²	CITB	1,664,208	10147.6
1951	69 km²	CITB	786,343	11396.3
1931	54 km²	British	309,786	5736.8
1541	2.64 km²	Kempegowda	11,841*	4485.2

* estimate

BBMP: Bruhat Bengaluru Mahanagara Palike
BDA: Bangalore Development Agency
CITB: City Improvement Trust Board

FIGURE 8.3
Growth of Bengaluru.

industrial and commercial establishments. These developments in the city and its outskirts provided job opportunities but also led to a population boom (Figure 8.3) with a demand for infrastructure and basic amenities. The shift of planned layouts (by BDA) to the development by private players led to unrealistic urban growth with erosion of open spaces including tree cover (Bharath et al., 2013).

8.3.1 Evolution of Bengaluru: From a Hamlet to Humongous Metropolis

The foundation stone for Bengaluru was laid during the era of Kempegowda (1537). During this time, Bengaluru was known as Bendakaluru (Land of Boiled Beans) with an area of about 2.64 sq km (Figure 8.4). Aadilshah of Bijapur captured Bengaluru during 1638 and expanded the area to

Bendakaluru (1537) mapped on 1731

Bangalore (1935)

FIGURE 8.4
Historical Bengaluru.

7.9 sq km before the British took over during the 1800s. During the British era, a cantonment and defence base were established in Bengaluru. By the end of British regime, Bendakaluru had become Bangalore with an area of 69 sq km. After independence, Karnataka state (known as the State of Mysore until 1973) was formed and during this period the CITB was formed. CITB increased the area from 69 sq km (1956) to 134 sq km (1976) with the setting up of government institutions (Public Sector Units) such as HAL, BEML, BHEL, BEL, ITI and HMT. The Bangalore Development Agency (BDA) was formed in 1976, and under the aegis of BDA, Bengaluru expanded with residential areas followed by large commercial establishments such as Peenya and Rajajinagar Industrial Area with the spatial extent of 161 sq km. IT hubs (SEZs) such as Electronic City in the early 2000s increased the city area to 226 sq km. Later, Bangalore became Bruhat Bengaluru (BBMP) in 2007; with the inclusion of 110 surrounding villages, the total area increased to 741 sq km.

Development in Bengaluru is currently driven by various agencies as explained in Table 8.1.

8.3.2 Status of Lakes and Trees in Bengaluru

Wetlands/lakes constitute the earth's most important freshwater resource, supporting huge biological diversity and providing a wide range of ecosystem services. Wetlands are highly productive ecosystems as they function as ecotones, transition zones between different habitats, and have characteristics of both aquatic and terrestrial ecosystems (DeGroot et al., 2006). Wetlands are also the most threatened and fragile ecosystems that are susceptible to changes owing to changes in the composition of their biotic and abiotic factors (Ramachandra & Bharath, 2015). They help in maintaining the ecological balance of the region and meet the need for life on Earth such as a source of drinking water, fish production, storage of water, sediment trapping, nutrient retention and removal, groundwater recharge and discharge, flood and erosion control, transport, recreation, climate stabilisers, support for food chains, and habitat for indigenous and migratory birds. Wetlands/ lakes also play a major role in treating and detoxifying a variety of waste products. Wetlands aid in remediation and aid as kidneys of landscape (Ramachandra et al., 2017b).

Wetland loss and degradation are due to conversion of wetland to non-wetland areas, encroachment of drainages (raja kaluves) through land filling, pollution due to sustained discharge of untreated domestic sewage as well as industrial effluents and dumping of solid waste, hydrological alterations (water withdrawal and inflow changes), and overexploitation of their natural resources. These anthropogenic activities result in habitat degradation, weed infestation due to nutrient enrichment, loss of biodiversity, and decline in goods and services provided by wetlands. Pollution of waterbodies is due to (Ramachandra et al., 2017a; Ramachandra et al., 2018):

TABLE 8.1

Agencies and Their Functions

Agency	Role
BDA: Bangalore Development Agency	• Land use planning, zoning • Creating and maintaining urban infrastructure • Assessment and collection of taxes • Monitors the execution of water supply and underground drainage works in BDA layouts taken up by BWSSB and electrification works executed by BESCOM The growth of any city is bounded by the city development plans laid by the BDA.
BBMP: Bruhat Bangalore Mahanagara Palike	• Creating and maintaining urban infrastructure • Assessment and collection of taxes Works similar to BDA
BWSSB: Bangalore Water Supply and Sewerage Board	• Distribution of water to the end users • Maintenance of the water supply and sanitary mains • Treatment of sewage (STP) • Permissions for development based on resource availability • Investigation of additional water demands for domestic purposes
BESCOM: Bangalore Electricity Supply Company	• Distribution of electricity to the end users • Maintenance of the supply mains, transformers, etc. • Perspective planning to meet the future demands
BMTC: Bangalore Metropolitan Transport Corporation	• Providing road transportation service to the public • Ensuring last mile connectivity
KIADB: Karnataka Industrial Area Development Board	• Acquire land and form industrial areas in the state • Provide basic infrastructure in industrial areas • Acquire land for single-unit complexes • Acquire land for government agencies for their schemes and infrastructure projects
KSPCB: Karnataka State Pollution Control Board	• Stipulating and auditing pollution control norms for industries, apartments • Enforcing rules and notifications of the Environment Protection Act • Plan programme for prevention and control/abatement of pollution of water, air and land
KFD: Karnataka Forest Department	Maintenance of green spaces (avenue trees, parks, etc.) Clearing of fallen trees and protection of trees
BMRCL: Bangalore Metro Rail Corporation limited	Providing metro rail service to the public, thus ensuring reduction in greenhouse gas emissions

1. Pollutants entering from point sources – Nutrients from wastewater from municipal and domestic effluents; organic, inorganic and toxic pollution from industrial effluents; storm-water runoff

2. Pollutants from non-point sources – Nutrients through fertilisers, toxic pesticides, etc., from agriculture runoff; organic pollution from human settlements near lakes/freshwater resources

The entry of untreated sewage into lakes has resulted in nutrient enrichment leading to eutrophication with algal blooms and macrophyte cover with dissolved oxygen depletion and malodour generation. Land use and land cover (LULC) changes in the wetland catchment alter the physical and chemical integrity of the system, which influences the biological community structure of the area.

Wetlands of Bengaluru occupy about 4.8% of the city's geographical area covering both urban and non-urban areas. Bengaluru has many manmade wetlands but no natural wetlands. In 1973, Bangalore (in the spatial extent of 221 sq km) had 207 waterbodies, which has now reduced to 93 (by 2011) and many lakes (54%) were either converted to layouts or encroached for illegal buildings. Field investigations reveal that nearly 66% of lakes are sewage fed, 14% surrounded by slums and 72% showed loss of catchment area. About 30% of the lakes were drained for residential sectors. About 22% of lakes had land filling and construction activities. Now, lake beds are being used as dumping yards for either municipal solid waste or building debris. The storm-water drains, lake beds, floodplains and catchment areas have been encroached and converted to layouts or for commercial purposes. Land use analyses reveal a 1028% increase in built-up areas with a decline of 88% vegetation and 79% waterbodies.

Frequent flooding (since 2000, even during normal rainfall) in Bengaluru is a consequence of the increase in impervious areas with the high-density urban development in the catchment and loss of wetlands and vegetation. This is coupled with narrowing and concretising storm-water drains, lack of appropriate drainage maintenance works with the changes in enhanced run-offs, the encroachment and filling in the floodplain on the waterways, obstruction by the sewer pipes and manholes and relevant structures, deposits of building materials and solid wastes with subsequent blockage of the system, and also flow restrictions from undercapacity road crossings (bridge and culverts). The lack of planning and enforcement has resulted in significant narrowing of the waterways and filling in of the floodplain by illegal developments.

Quantification of number of trees in the region using remote sensing data (of spatial resolution 5.8 m) with field census reveals that there are only 1.5 million trees to support Bengaluru's population of 9.5 million, indicating one tree for every seven persons in the city (Ramachandra & Bharath, 2016; Bharath et al., 2018b; Ramachandra et al., 2017). This is insufficient even to sequester respiratory carbon (ranges from 540–900 g per person per day).

8.4 SDSS Framework for Urban Planning

Schema of the spatial decision support system for visualising spatio-temporal growth patterns of Bengaluru and planning sustainable growth in the

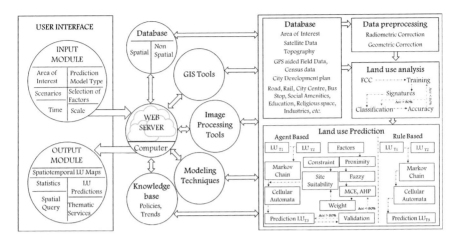

FIGURE 8.5
Schema of SDSS for Bengaluru.

city is presented in Figure 8.5. The SDSS framework consists of three components, namely (1) graphic user interface (GUI), (2) web server (with WMS, WFS options); and (3) system segment consisting of knowledge base, database, database management, simulation and geovisualisation modules.

The user interface is a graphical front end that allows the decision maker to provide various input options and derive outputs. In the input module, the user/decision maker can provide various information (as explained in Table 8.2). Based on the input, the web server will communicate and interact with the system segments to derive appropriate information and produce outputs as described in Table 8.3. The web server interacts with the user interface and the system segments. Based on the user input, the web server connects with various spatial and attribute databases to fetch information; the web server connects with other tools to derive necessary outputs and display them on the user interface.

The system segment in the current SDSS consists of five elements: the database management module, GIS, image processing module, modelling module, and knowledge base. The description of each is presented in Table 8.4. The back end working principles of various system segment tools are described in Figure 8.5.

The spatial decision support system helps in understanding the possible future growth given various criteria. In the current article, likely land uses due to urbanisation of Bengaluru are simulated through statistics regression models and spatial models. Regression models consider existing growth patterns and population trends. Spatial models with two different scenarios have been designed as explained in Table 8.5.

Land use dynamics in Bengaluru during 2016 and 2019 are depicted in Figure 8.6. Long-term analysis shows that the urban area has increased from 7.79% (in 1973) to 81.05% (in 2019) with decrease in green spaces from

TABLE 8.2

Description of Various Input Variables

Input Variable	Description
Area of interest	• The user can provide the details on area he/she is working/planning. • The input can be spatial or attribute. • Spatial input can be a shape file or any other vector file format. • Attribute input allows the user to select the boundaries by selecting pre-defined variables (district/taluk, ward/village, etc.). Based on this input the area of interest would be selected.
Time step	• The user can provide the initial time, end time and time steps at which the land use needs to be derived from satellite data and predicted for the future.
Scale	• Scale of work depends on area of interest, i.e., village, panchayat, taluk, district, state, etc. • With scale of work can be suited with the resolution of satellite data. Example: 30 m satellite data for scale of 1:50000.
Scenarios	• Users can visualise land use changes/urbanisation considering various scenarios by providing criteria information such as the business as usual scenario, protection scenario, or sustained growth scenario.
Prediction model	• The predictive models can be chosen as either rule based or agent based. This depends also on the scenario considered.
Selection of factors	• Various factors (causal or constrains) driving the change are either set based on the scenario selected or can be manually selected for visualising the future scenarios. • Users can also prioritise the causal factors based on expert judgement.

68.27% (1973) to 3.98% (2019) and decline in waterbodies from 3.4% (1973) to 1.75% (2019). The developmental trends within the administrative area show affinity towards concentrated urban growth. Comparative analyses of current growth patterns with the CDP of 2015 (Figure 8.7) highlight deviation from the planned growth. Drastic increase in urban area (paved surfaces)

TABLE 8.3

Description of Various Output Variables

Input Variable	Description
Land use	• The decision maker/user can visualise the present status of land use and spatial patterns of land use dynamics.
Statistics	• Land use statistics across select time are returned as output. • Land use changes from one class to another class.
Spatial query	• The decision maker can query the spatial data in terms of changes in land use at various levels (pixel/AOI). • Persistency of land use at different time scales.
Land use predictions	• Future land use trends based on different scenarios or predictive models are depicted showing possible changes.
Thematic services	• Various thematic maps such as land use, topography, factors and data used with sources are produced for better interpretation and printable formats for reports.

TABLE 8.4

System Segments

Segment	Description
Database	• Spatial and attribute databases are stored. • Databases involve road networks, satellite data, classified land use (ancillary), topographic data, census data, educational institutes, industries, bus stops, field data with photographs (dated), comprehensive CPD, etc.
GIS	• Various open source and programming tools such as QGIS (available at www.qgis.org) are used in the back end to analyse, query and visualise the data.
Image processing tool	• An image processing tool (GRASS GIS, http://grass.osgeo.org) is used to correct satellite data for geometric and radiometric errors and classify the image (only if not available in the database).
Modelling/ simulation module	• Predicting future landscapes based on various criteria and scenarios provided by the user using agent-based or rule-based models. • Agent-based models incorporate all the factors and constrains, while rule-based models use the current context only.
Knowledge base	• This includes the policies and rules laid by the BDA and other authorities (e.g., buffer zones for protecting water bodies). • The knowledge base will also have historical data information (same as database). • Earlier trends, resource availability, etc., are a part of the knowledge base.

has impacted the groundwater table evident from the depth beyond 500 m and green cover. Sustained inflow of untreated or partially treated sewage and industrial effluents have impacted both surface water and groundwater resources. The landscape integrity is compromised with the encroachments of (1) lakes, (2) storm-water drains, (3) valley zones (low-lying areas connecting lakes), (4) floodplains/buffer zones, and (5) green cover or open spaces. Impacts are evident from the frequent occurrences of floods, lowered groundwater table, inability of the catchment to retain water, lowered oxygen levels, etc. This has necessitated understanding the land use dynamics with the likely changes in the future considering various scenarios, which would help in decision making towards the sustainable management of natural resources.

Long-term urban growth trends considering 1973 to 2019 confirm the sharp increase in paved surface with the decline of waterbodies as well as green spaces. In recent years, the spatial patterns of urbanisation within the BBMP limits highlight the near saturation of landscape with paved surfaces (post 2016).

8.4.1 Scenario 1: Business as Usual Scenario

The predicted landscape for scenario 1 is depicted in Figure 8.8. Land use for the year 2016 and 2019 were used to simulate 2019 (which was validated with the actual land uses in 2019) and predict for the years 2022, 2025,

TABLE 8.5

Scenarios Designed for Decision Support

Model	Growth Factors	Constraints
Rule-based model	None	Existing waterbodies
Description:	• The scenario is also knows as the business as usual scenario. • The model uses cellular automata (CA) and Markov chains to predict the future landscapes in BBMP area. • Markov chains give the change probability for one land use type to the other, and CA spatially distributes the possible areas of change based on rules: (1) urban area will not change to any other land use; (2) vegetation, water and others can change to urban; and (3) vegetation can change to other land use categories.	
2. Agent-based model	• Road network • Railway and bus shelters • Educational institutions • Industries • Socio-cultural and religious spaces • Metro	• Protected areas as per CDP 2031 • Defence areas • Institutions (IISc, GKVK, BU) • State forests • Valley zone • Buffer zones • Slope • Waterbodies
Description:	• The scenario is also knows as the sustained growth scenario. • This scenario considers various constraints and causal factors of growth. • It helps in understanding the effective distance of influence of various causal factors for different land uses using overlay operations and histograms. • These distances are normalised using fuzzy logic. • The relative influence (weights) of each of the factors is determined using the multi-criteria evaluation technique that depends on expert judgement. • Weights are used to derive site suitability maps using the analytical hierarchical process for various land use types. • These site suitability maps are used along with CA and Markov chains to predict future land use in BBMP.	

2028 and 2031. Comparison of actual land uses with the simulated for 2019 showed 94% accuracy. On confirmation of robustness of the model, it is used to predict for the year 2031 with 3-year time steps. Prediction results show that the built-up growth saturates Bengaluru by 2028, covering an area of 90% with concentrated growth, while vegetation will reduce to 2.1% and others land use will reduce to 5.3%.

8.4.2 Scenario 2: Sustained Growth Scenario

In the sustained growth scenario, various factors contributing to growth either causal or constraints were used as tabulated in Table 8.5. Figure 8.9 shows some of the factors considered for predicting future landscape changes in the study area. Slope and protected areas such as water bodies, defence, green spaces, institution campuses (Indian Institute of Science

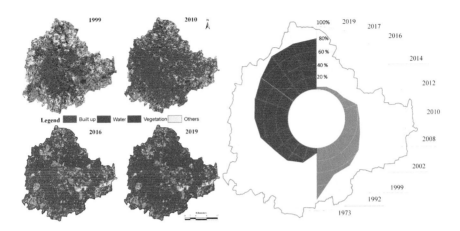

FIGURE 8.6
Trends in landscape dynamics of Bengaluru (BBMP).

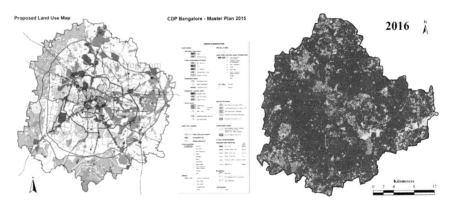

FIGURE 8.7
CDP 2015 and land use 2016.

[IISc]; University of Agricultural Sciences, Gandhi Krishi Vigyana Kendra [UAS, GKVK]; Bangalore University [BU]) act as a constraints, while roads, education institutes, industries, social amenities, bus and railway stations act as major causal factors of change. Effective distances for each of the causal factors with urbanising pockets are estimated.

For example, the proximity map (buffer) of road is overlaid on built-up area, and the histogram is plotted to derive the effective ranges of influence (Figure 8.10). Distance to roads and built-up have a negative sigmoidal relationship, i.e., the influence of road on urbanisation starts decreasing as built-up transits away from the main road. In the current example, roads have the highest influence until 200 m and influence starts declining between 200 and 1500 m, and there are no built-up areas after 2000 m from the road.

The influence distances are used to normalise (between 0 [no influence] and 255 [high influence]) using fuzzy logic and constrains are converted

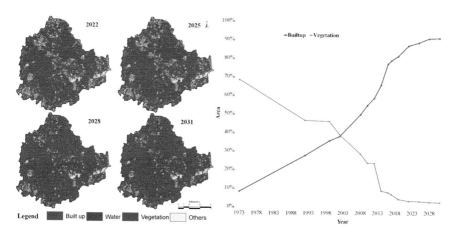

FIGURE 8.8
Land use prediction using business as usual scenario.

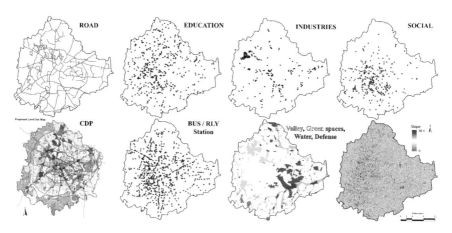

FIGURE 8.9
Agents of change.

to Boolean (0 – no change, 1 – changes can occur). These normalised data are used to derive the site suitability maps based on which simulation is carried out. Site suitability maps for various land use classes are depicted in Figure 8.11.

Land use of 2016 was used to simulate 2019 considering the site suitability and transition conditions. The simulation results of 2019 were closer to the actuals of 2019 with an accuracy of 96%. This calibrated model was further used to predict for the year 2031 with 3-year time step.

Land use prediction using the sustained growth (Figure 8.12) scenario shows that Bengaluru landscape would be saturated by 2025 with built-up areas covering about 82.9%; vegetation in the city will drop to 3.4% and others

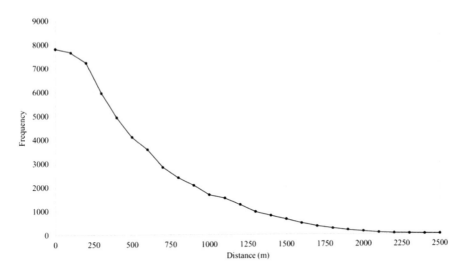

FIGURE 8.10
Effective influencing distance of roads on built-up.

to 12%. The regions which are considered as constraints (assumed to continue to be in the same land use in spatial models) are depicted in Figure 8.13, which include campuses of educational institutions (IISc, GKVK, BU, etc.), state forests, defence establishments (land army, air force, etc.), parks (Lalbagh, Cubbon Park, J P Park, Freedom Park) and government establishments (office

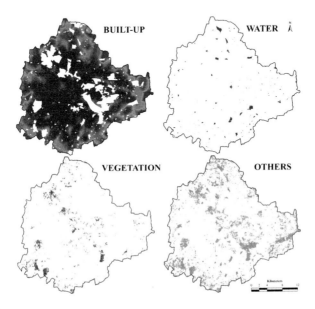

FIGURE 8.11
Site suitability maps.

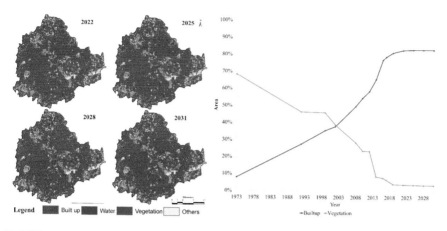

FIGURE 8.12
Land use prediction using sustained growth scenario.

spaces at Malleshwaram, Karnataka Forest Department, Karnataka High Court Complex, Vidana Souda, Vikasa Souda, etc.).

In addition to the spatial models, statistical modelling (regression analysis) was carried out to understand the growth rates. Linear regression was carried out between proportionate built-up (P_b) and time of the last decade (2008 to 2019). Equation 8.1 explains the likely growth of Bengaluru city. In the regression analysis no constrains were considered, i.e., all land uses can

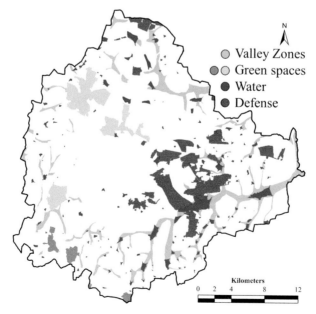

FIGURE 8.13
Regions which are considered as constraints in spatial agent based models.

change to urban area. Prediction based on the regression analysis shows that the Bengaluru landscape would be saturated or completely urbanised by the year 2025, with no spaces for any other land uses.

$$P_b = 63.814 * \ln(y) - 484.82,$$
$$R^2 = 0.968 \text{ for } P_b < 100\%$$
$$P_b = 100\% \text{ for } P_b > 100\% \tag{8.1}$$

where P_b is the proportionate built-up area, y is the year, and R^2 is the coefficient of determination.

Similarly, multiple regression between population, urban growth and time (Equation 8.2) shows that Bengaluru would be completely urbanised by 2023:

$$P_b = 0.02617 * (y) + 1.045/100000 * (PD_y) - 52.2011,$$
$$R^2 = 0.984 \text{ for } P_b < 100\%$$
$$P_b = 100\% \text{ for } P_b > 100\% \tag{8.2}$$

where P_b is the proportionate built-up area, y is the year, Pop_y is the alpha population density during the year, and R^2 is the coefficient of determination.

Comparing the spatial growth trends (Figure 8.14) in the business as usual (RBM) and sustained growth (ABM) scenarios, RBM shows saturation by 2028 with the city getting urbanised to 90.8%, while ABM, considering all the constraints indicated, shows the city gets saturated by 2025 with urban

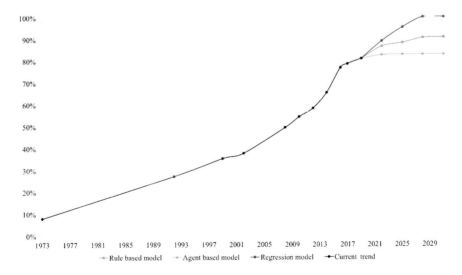

FIGURE 8.14
Comparative analysis of urban growth patterns.

areas covering 82.9% of BBMP. ABM supports higher vegetation cover with open spaces, i.e., about 3.4% and 12%, respectively, as against RBM with 2.1% and 5.3%, respectively, by 2031. The regression model shows that by 2025 the entire area would be urbanised. This indicates the importance of sustained growth models for conserving the green cover and open spaces in the city, which would further help in moderating air quality, micro climate (temperature) and hydrological regimes (surface and subsurface) at local scale.

The study shows that regression models, RBM and ABM predict differently, providing insights to various scenarios such as extreme unabated scenarios to conservation conditions. The scenario based results helps in the decision making to take appropriate policy measures for regulating growth (land use alteration) while catering to the basic amenities and infrastructure for the local dwellers.

8.5 Conclusions

A spatial decision support system with the database management system, simulation model and geovisualisation capability aids in the decision making towards prudent management of natural resources as the decision maker is equipped to visualise implications of the policy decisions and also likely growth with the current trends of urbanisation. SDSS provides the decision maker/user with three scenarios, i.e., an extreme urbanisation scenario to controlled and sustainable growth scenarios. Urban growth in Bengaluru is analysed to demonstrate the utility of SDSS in the regional planning.

Spatial temporal patterns analyses of urbanisation process in Bengaluru reveals that of gradual increase in paved surfaces during 1973 (8%) and 2002 (37.7%). However, the infrastructural boom due to the IT industries and large-scale SEZ in the peri-urban regions during 2000 witnessed transformation of peri-urban landscape. These regions experiencing sprawl are still devoid of basic amenities and infrastructure. Subsequently, urban areas increased to 76.9% (2016) and 81.05% (2019), with telling impacts on the local ecology and environment evident from the decline of waterbodies, agriculture (others) lands and green cover with the enhanced levels of pollutants in the environment. The vegetation cover declined from 67% (1973) to 3.8% (2019) and there has been a 79% loss of waterbodies. The likely changes in land use are predicted using various models such as the statistical model and spatial models. The results reveal saturation of the Bengaluru landscape with the current trends of urbanisation by 2025 (through regression models) and 2023 (based on current population dynamics).

Spatial models such as the rule-based CA-Markov model and agent-based fuzzy-AHP-CA models have a prediction accuracy >94%. Rule-based models show that the urban areas would be saturated covering an area of 90.5%

by 2028 (excluding waterbodies, parks and institution campuses) and will be persistent (90.7%) in 2031. Agent-based models considering causal factors with the constraints reveal the landscape getting saturated by 2025.

Insights of the causal factors of changes, growth limits, and environmental implications with the history of LULC changes aid in evolving prudent management strategies, revisiting the policies and plans (if required) to design sustainable and liveable cities.

Acknowledgements

We are grateful to (i) UNSD (United Nations Statistics Division); (ii) The Ministry of Statistics and Programme Implementation, GoI; (iii) ISRO-IISc Space Technology Cell, Indian Institute of Science and (iv) Indian Institute of Technology, Kharagpur (V) SERB, Department of Science and Technology for infrastructure and financial support. We thank (i) United States Geological Survey and (ii) National Remote Sensing Centre (NRSC, Hyderabad) for providing temporal remote sensing data.

References

Arsanjani, J. J., Helbich, M., Kainz, W., & Boloorani, A. D. 2013. Integration of logistic regression, Markov chain and cellular automata models to simulate urban expansion. *International Journal of Applied Earth Observation and Geoinformation*, 21, 265–275.

Bharath, H. A., Chandan, M. C., Vinay, S., & Ramachandra, T. V. 2018a. Urbanisation in India: Patterns, visualisation of cities and greenhouse gas inventory for developing an urban observatory. In Weng, Q., Quattrochi, D., & Gamba, P. E. (Eds.), *Urban remote sensing*, 2nd edition. Boca Raton: CRC Press; Taylor and Francis.

Bharath, H. A., Vinay, S., & Ramachandra, T. V. 2014. Prediction of spatial patterns of urban dynamics in Pune, India. In *2014 Annual IEEE India Conference (INDICON)* (pp. 1–6). IEEE.

Bharath, H. A., Vinay, S., & Ramachandra, T. V. 2016. Agent based modelling urban dynamics of Bhopal, India. *Journal of Settlements and Spatial Planning*, 7(1), 1.

Bharath, H. A., Vinay, S., Chandan, M. C., Gouri, B. A., & Ramachandra, T. V. 2018b. Green to gray: Silicon Valley of India. *Journal of Environmental Management*, 206, 1287–1295.

Bharath, H. A., Vinay, S., Durgappa, S., & Ramachandra, T. V. 2013. Modeling and simulation of urbanisation in greater Bangalore, India. In *Proceedings of National Spatial Data Infrastructure 2013 Conference*, IIT Bombay (pp. 34–50).

Burrough, P. A. 1986. *Principles of geographical information systems for land resources assessment.* Oxford University Press.

Census of India, Office of the Registrar General & Census Commissioner, Ministry of Home Affairs, Government of India. 2011. Accessed on February 10, 2019. Retrieved from http://www.censusindia.gov.in

Cerreta, M., & De Toro, P. 2012. Urbanization suitability maps: A dynamic spatial decision support system for sustainable land use. *Earth System Dynamics, 3*(2), 157–171.

Chen, S., Zeng, S., & Xie, C. 2000. Remote sensing and GIS for urban growth analysis in China. *Photogrammetric Engineering and Remote Sensing, 66*(5), 593–598.

Crossland, M. D. 2008. Spatial decision support system. In *Encyclopedia of GIS*, Eds., Shekhar, S., & Xiong, Z. (pp. 1095–1095). Boston, MA: Springer US.

Crossland, M. D., Wynne, B. E., & Perkins, W. C. 1995. Spatial decision support systems: An overview of technology and a test of efficacy. *Decision Support Systems, 14*(3), 219–235.

De Groot, R., Stuip, M., Finlayson, M., & Davidson, N. 2006. *Valuing wetlands: Guidance for valuing the benefits derived from wetland ecosystem services* (No. H039735). International Water Management Institute.

DESA/Population Division United Nations. 2018. *World urbanization prospects 2018.* Accessed on July 10, 2019. Retrieved from https://population.un.org/

Grekousis, G., Manetos, P., & Photis, Y. N. 2013. Modeling urban evolution using neural networks, fuzzy logic and GIS: The case of the Athens metropolitan area. *Cities, 30*, 193–203.

Grimm, N. B., Faeth, S. H., Golubiewski, N. E., Redman, C. L., Wu, J., et al. 2008. Global change and the ecology of cities. *Science, 319*(5864), 756–760.

He, J., Liu, Y., Yu, Y., Tang, W., Xiang, W., et al. 2013. A counterfactual scenario simulation approach for assessing the impact of farmland preservation policies on urban sprawl and food security in a major grain-producing area of China. *Applied Geography, 37*, 127–138.

Jain, R. K., Jain, K., & Ali, S. R. 2017. Modeling urban land cover growth dynamics based on land change modeler (LCM) using remote sensing: A case study of Gurgaon, India. *Advances in Computational Sciences and Technology, 10*(10), 2947–2961.

Ji, C. Y., Liu, Q., Sun, D., Wang, S., Lin, P., et al. 2001. Monitoring urban expansion with remote sensing in China. *International Journal of Remote Sensing, 22*(8), 1441–1455.

Mena, C. F., Walsh, S. J., Frizzelle, B. G., Xiaozheng, Y., & Malanson, G. P. 2011. Land use change on household farms in the Ecuadorian Amazon: Design and implementation of an agent-based model. *Applied Geography, 31*(1), 210–222.

Mohammady, S., Delavar, M. R., & Pahlavani, P. 2014. Urban growth modeling using AN artificial neural network a case study of Sanandaj City, Iran. *The International Archives of Photogrammetry, Remote Sensing and Spatial Information Sciences, 40*(2), 203.

NASA. 2001. *Satellite maps provide better urban sprawl insight.* NASA, News Release 2, 2001. Accessed on January 23, 2019. Retrieved from https://www.nasa.gov/

Rafiee, R., Mahiny, A. S., Khorasani, N., Darvishsefat, A. A., & Danekar, A. 2009. Simulating urban growth in Mashad City, Iran through the SLEUTH model (UGM). *Cities, 26*(1), 19–26.

Ramachandra, T. V. 2009. RIEP: Regional integrated energy plan. *Renewable and Sustainable Energy Reviews, 13*(2), 285–317.

Ramachandra, T. V., & Bharath, H. A. 2012. Land use dynamics at Padubidri, Udupi District with the implementation of large scale thermal power project. *International Journal of Earth Sciences and Engineering*, 5, 409–417.

Ramachandra, T. V., & Bharath, H. A. 2015. Wetlands: Kidneys of Bangalore's landscape. *National Wetlands*, 37, 12–16.

Ramachandra, T. V., & Bharath, H. A. 2016. Bengaluru's reality: Towards unlivable status with unplanned urban trajectory. *Current Science*, 110(12), 2207–2208.

Ramachandra, T. V., Bharath, H. A., Kulkarni, G., & Vinay, S. 2017a. Green spaces in Bengaluru: Quantification through geospatial techniques. *Indian Forester*, 143(4), 307–320.

Ramachandra, T. V., Bharath, H. A., & Sanna, D. D. 2012. Insights to urban dynamics through landscape spatial pattern analysis. *International Journal of Applied Earth Observation and Geoinformation*, 18, 329–343.

Ramachandra, T. V., Krishna, S. V., & Shruthi, B. V. 2005. Decision support system for regional domestic energy planning. *Journal of Scientific and Industrial Research*, 64, 163–174.

Ramachandra, T. V., Kumar, R., Jha, S., Vamsee, K., & Shruthi, B. V. 2004. Spatial decision support system for assessing micro, mini and small hydel potential. *Journal of Applied Sciences*, 4(4), 596–604.

Ramachandra, T. V., Mahapatra, D. M., Vinay, S., Varghese, S., Asulabha, K. S., et al. 2017b. *Bellandur and Varthur Lakes rejuvenation blueprint*. ENVIS technical report 116. Bangalore: Environmental Information System, CES, Indian Institute of Science.

Ramachandra, T. V., Sudarshan, P. B., Mahesh, M. K., & Vinay, S. 2018. Spatial patterns of heavy metal accumulation in sediments and macrophytes of Bellandur wetland, Bangalore. *Journal of Environmental Management*, 206, 1204–1210.

Ramachandra, T. V., Tara, N. M., & Bharath, S. 2017c. Web based spatial decision support system for sustenance of western ghats biodiversity, ecology and hydrology. In Aneesha, S., & Jamuna, R. (Eds.), *Creativity and cognition in art and design* (pp. 58–70), 1st edition. Bangalore: Bloomsbury.

Siddayao, G. P., Valdez, S. E., & Fernandez, P. L. 2014. Analytic hierarchy process (AHP) in spatial modeling for floodplain risk assessment. *International Journal of Machine Learning and Computing*, 4(5), 450.

Sugumaran, R., & Degroote, J. 2010. *Spatial decision support systems: Principles and practices*. CRC Press.

Sun, S., Parker, D. C., Huang, Q., Filatova, T., Robinson, D. T., et al. 2014. Market impacts on land-use change: An agent-based experiment. *Annals of the Association of American Geographers*, 104(3), 460–484.

Taubenböck, H., Esch, T., Felbier, A., Wiesner, M., Roth, A., et al. 2012. Monitoring urbanization in mega cities from space. *Remote Sensing of Environment*, 117, 162–176.

Taubenböck, H., Wegmann, M., Roth, A., Mehl, H., & Dech, S. 2009. Urbanization in India–Spatiotemporal analysis using remote sensing data. *Computers, Environment and Urban Systems*, 33(3), 179–188.

TERI. 2019. Urbanisation, decoding urban climate change resilience: The policy way. Accessed on July 15, 2019. Retrieved from https://www.teriin.org/

Vinay, S., Bharath, S., Bharath, H. A., & Ramachandra, T. V. 2013. Hydrologic model with landscape dynamics for drought monitoring. In *Proceeding of: Joint International Workshop of ISPRS WG VIII/1 and WG IV/4 on Geospatial Data for Disaster and Risk Reduction*, Hyderabad, November (pp. 21–22).

Voigtländer, S., Breckenkamp, J., & Razum, O. 2008. Urbanization in developing countries: Trends, health consequences and challenges. *Journal of Health and Development*, 4(1–4), 135–163.

Wang, H., He, S., Liu, X., Dai, L., Pan, P., et al. 2013. Simulating urban expansion using a cloud-based cellular automata model: A case study of Jiangxia, Wuhan, China. *Landscape and Urban Planning*, *110*, 99–112.44. Zadeh, L. H., 1965. Fuzzy Sets, *Information and Control, 8*, 338–353.

Index

A

Accuracy
  assessment, 8, 61, 65
  overall, 8, 66, 67, 91
  producer's, 8, 66
  user's, 8, 66
Adaptive capacity, 127–129,
      131–133, 139
Advection, 111
Aerosols, 22, 110
Agent Based Model (ABM), 9, 10, 85, 92,
      151, 161, 169
Aggregation index, 71
Air quality, 25, 30, 35
Albedo, 107, 109–111
Altitude, 5, 111
Analytical Hierarchical Process (AHP),
      81, 91, 144, 151
Angle of incidence, 110
Annual Average Rate of Change
      (AARC), 8
Annual Increase Index (AII), 8
Annual Land Use Growth Rate Index
      (ALGRI), 8
Anthropogenic activity, 2, 39, 41, 45,
      134, 137, 156
Anthropogenic, 21–24, 27, 29, 31, 32,
      108, 119
Arable land, 41
Artificial intelligence, 9, 10, 55, 85, 102, 152
Atmospheric correction, 119

B

Bhuvan, 52, 96
Biodiversity, 22, 107
Biological Oxygen Demand
      (BOD), 28
Black body, 110–113
Bruhat Bangalore Mahanagara Palike
      (BBMP), 153, 154, 156–157
Buoyancy, 111

C

Calibration
  coarse, 96, 97
  final, 96, 97, 99
  fine, 96, 97
  process, 95
Calorific value, 30
Carbon
  content factor, 32
  decomposable fraction, 32
  degradable organic, 32
  intensity, 25
  oxidation factor, 32
Carbon dioxide equivalent (CO_2e), 20,
      24, 27, 29, 34
Catchment area, 130, 134–135, 158
Cellular automata, 9, 85, 86–93, 151, 162
Change detection, 3, 39
Chemical Oxygen Demand (COD), 28
Chennai Metropolitan Area (CMA), 52
City Development Plan (CDP), 49, 64, 92,
      96, 153, 160, 162
Classification, 7, 39, 43, 48, 51, 65
Climate
  change, 22, 23, 25, 26, 45, 83, 107, 133
  macro, 19
  micro, 19, 35, 109
Climatology, 108, 117
Cloud cover, 110, 119
Conduction, 110
Confusion matrix, 8, 66
Continental precipitation, 22
Conurbations, 128
Convection, 110, 111
Coordinate values, 3

D

Data
  acquisition, 84
  attribute, 3, 152
Database management subsystem, 153

Database Management System
        (DBMS), 3, 46
Decarburisation, 25
Decision support systems, 4
Deforestation, 41, 44–46, 62
Denitrification, 22, 32
Density gradient, 45, 63
Developing countries, 5, 15, 41, 43, 45, 55
Digital Elevation Model (DEM), 96, 138
Disaster management, 129, 139
Dissolved oxygen, 158

E

Early warning system, 129
Earth observation, 9, 47
Earth's atmosphere, 3
Ecology, 108, 134, 149, 150
Economic development, 25
Effluents
    domestic, 157
    industrial, 157, 161
    municipal, 157
Electromagnetic spectrum, 48
    long-wave, 19, 108, 109, 113, 119
    shortwave, 19, 108
    thermal infrared, 19, 113–115, 117
    ultraviolet, 19
    visible, 19, 117
Elitism, 99
Emissions
    respiration, 21
    ocean release, 21
    deforestation, 21
    natural oil, 21
    manure production, 21
    wildfires, 21
    wetlands, 21
    refrigeration and air conditioning, 22
    inventory, 26
    sector-wise, 26, 29, 30, 34
    particulate matter, 27, 28, 30
    fly ash, 27, 30
    mineralisation, 32
    industrial process and production,
        21, 22
    biomass burning, 21, 22, 27
    fossil fuels, 21, 22, 30
    permafrost, 21, 23

waste management, 21, 26, 29
decomposition, 21, 28
cultivation of rice/paddy, 21, 28
enteric fermentation in cattle, 21, 29,
        33, 34
coal, 21, 30
land use change, 21, 35, 107
Emission factor, 29–33
Emissivity, 107, 110–112, 114, 119–122
Encroachment, 135, 137
    catchment area, 134
    drainages, 156
    ecologically sensitive area, 134
    lake, 135, 161
    land, 91
    wetland, 45
Energy
    budget, 108, 109
    efficiency, 25, 26
    renewable, 25, 26
Energy interactions, 3
Energy source, 3
Environmental cycles
    bio-geo-chemical cycles, 107
    carbon cycle, 107
    nitrogen cycle, 107
    water cycle, 107
Environmental models
    global climatic capriciousness, 108
    ocean circulation models, 108
    weather prediction, 108
Environmental-related treaties
    Kyoto protocol, 23
    Rio convention, 23
Error of commission, 8
Error of omission, 8
Eutrophication, 158
Evapotranspiration, 22, 108
Excluded map, 96

F

False Colour Composite (FCC), 5, 6,
        52, 53
Field data, 52
Flood plain, 158
Flood vulnerability zoning, 134,
        136–137
Forest cover change, 41, 49

global, 43
Fragstat, 63
Framework, 25, 26, 35
Future scenario, 83, 141, 151, 160
Fuzzy
    logic, 91, 150, 162–163
    rules, 151–152

G

Gaussian Maximum Likelihood
        Classifier (GMLC), 7, 67, 141
Genetic algorithm, 95, 99, 101, 151
Geographic Information System (GIS), 2,
        40, 46, 48, 90, 150
Geo-informatics, 2, 3
Georeferencing, 65
Geospatial
    data, 2, 3, 4
    technology, 48, 51, 134
Gigalopolis, 94, 95
Global Navigation Satellite System
        (GNSS), 4
Global Positioning System (GPS), 3, 52
Global Warming Potential (GWP), 21, 22,
        24, 30
Global warming, 22–24, 122
Globalisation, 128, 130, 134, 149
Goodness of fit, 95, 96
Google earth, 53
Google maps, 96
Gradient analysis, 69, 76
Graphic User Interface (GUI), 159
Greenhouse effect, 19, 20
Greenhouse Gases (GHGs)
    carbon dioxide (CO_2), 20–22, 24, 25,
        27, 28, 30–32
    ozone, 19–21
    chlorofluorocarbons, 20
    hydrofluorocarbons, 20
    footprint, 19, 21, 23–26, 29, 33
    water vapour, 19, 21, 120, 121
    sulphur hexafluoride, 19, 22, 24
    nitrogen, 19, 28, 33, 107
    trace gases, 20
    methane, 19, 21, 24, 28, 29, 32, 33
Grids, 3
Gross Domestic Product (GDP), 25, 28,
        34, 64

Ground Control Point (GCP), 52
Ground stations, 3
Ground truth, 4
Ground water table, 161

H

Habitat degradation, 156
Habitat reservoirs, 62
Hazard, 128, 129, 131, 133, 140
Heat flux, 108
High resolution, 6

I

Image analysis, 51
Image fusion, 51
Image processing, 159, 161
Impervious area, 158
India, 5, 45, 49, 54, 91, 98, 154
Indian Regional Navigation Satellite
        System (IRNSS), 4
Industrial effluents, 156
Industries
    aluminium, 28
    cement, 28, 31
    chemical, 28, 31
    glass, 28
    iron and steel, 28, 31
    paper and pulp, 28
    petrochemical, 28, 31
Industrialisation, 19, 28
Infrastructure, 25, 108
Instantaneous Field of View
        (IFOV), 5
Intergovernmental Panel for Climate
        Change (IPCC), 19, 20, 22–24,
        26, 30–32
International Geosphere-Biosphere
        Program (IGBP), 10, 81
Interspersion and juxtaposition,
        71, 76
Invasive species, 54, 82
Irradiance, 112

K

Kappa coefficient, 8, 66, 91
Kirchhoff's law, 112

L

Lambertian, 112, 113
Landscape, 1, 62, 108, 114
  changes, 5, 15, 162
  dynamics, 2, 4, 5, 55, 61, 87, 150
  fragmentation, 1, 5, 45
  integrity, 161
  metrics, 8, 62
  modelling, 81
  monitoring, 5, 6
  planning, 5, 63
  shape index, 62
  structure, 1, 5, 8, 15, 63, 82
Land cover, 1, 40
  change, 40
Land Surface Temperature (LST)
    algorithms
  radiative transfer, 121
  single-channel, 119, 120
  split-window, 118, 120, 121
Land use, 1, 40
  analyses, 141
  assessment, 6
  change, 10, 12, 39, 40, 55, 90, 151
  classification, 141, 144
  dynamics, 62, 65, 68, 77, 82, 86, 93,
    159–161
  modelling, 9
Land Use Expansion Intensity Index
    (LEII), 8
Land Use Land Cover (LULC), 39, 48, 81
  analysis, 84
Land Use Land Cover Change (LULCC),
    2, 4, 6, 7, 13, 39, 40, 41, 43, 44, 46,
    51, 55, 81, 82, 93, 102, 158
  dynamics, 2, 62
  model, 10
Largest patch index, 71, 76
Lee-Sallee metrics, 96, 97
Likert scale, 144
Line of sight, 113
Linear regression, 166
Liveability, 129, 131–132

M

Machine learning, 9, 39, 51, 55
Macrophyte, 158

Marginal worker, 141, 146
Markov chain, 10, 85, 89, 92, 151, 162
Megacities, 1, 5, 13, 64
Membership function, 151
Methane correction factor, 32
Millennium Development Goal
    (MDG), 128
Mitigation, 29, 107, 127, 129, 131, 136,
    139–140
Model management subsystem, 153
MODTRAN, 121
Moisture, 107, 111
Monte Carlo iterations, 94, 96
Moore neighbourhood, 88, 93
Multi Criteria Evaluation (MCE), 89, 91,
    137, 151
Multi resolution, 2, 47, 48, 114, 115, 150
Multi spectral, 5, 47, 52
Multiple internal reflection, 108

N

National Aeronautics and Space
    Administration (NASA), 47
Natural ecosystem, 42, 55, 86, 98
Natural resource, 2, 40, 45, 61, 82,
    161, 168
Near infrared (NIR), 48, 52
Nitrification, 22, 32
NOAA (National Oceanic
    and Atmospheric
    Administration), 19, 21, 23,
    115–116, 118, 121
Non-vegetation, 2
Normalised Difference
    Vegetation Index (NDVI),
    48, 49, 110, 137
Normalised Landscape Shape Index
    (NLSI), 69, 70, 73
Number of patches, 62, 69, 70, 72
Nutrient enrichment, 156, 158
Nutrient retention, 156

O

Ocean acidification, 22
Open source, 3, 94–95
Operational Land Imager, 47, 115
Optimal SLEUTH Metric (OSM), 94

P

Panchromatic, 117
Particulate matter, 27, 28, 30
Parts per billion (ppb), 21, 22
Parts per million (ppm), 21
Patch density, 71, 74
Patterns
    crop, 107
    wind, 107
Picture elements, 3
Pixel, 3, 5, 144
Planck's function, 111, 112
Policies, 25, 26, 29
Pollutants
    non-point sources, 157
    point sources, 157
Population initialisation, 99
Population, 23, 28, 30, 33, 108
Precipitation, 22, 108
Preparedness, 134
Pre-processing, 84
Principal component analysis, 43
Proximity map, 163
p-SLEUTH, 95

R

Radar, 51
Radiation, 19, 108–114, 118, 119
    down-welling, 109, 112, 114
    up-welling, 109, 112, 114
Radiative cooling, 108
Resolution
    radiometric, 40, 47
    spatial, 115–118, 121
    temporal, 5, 40, 47, 116, 117, 119
Raster data model, 3, 4
Raster image, 5
Reference data, 4
Regional planning, 2
Remote sensing, 2, 5, 15, 40, 41, 43,
        46, 49, 55, 110, 114, 115, 119,
        136, 144
    active, 3
    data, 2, 8, 39, 48, 61, 63, 76,
        136, 158
    optical, 47
    passive, 3

Resilience, 129, 131–134, 139
Rise in temperatures
    droughts and floods, 22
    infectious diseases, 22
    thermal discomfort, 22, 108
    thermal instability, 22, 121
    weather extremes, 22
Risk, 131, 134, 136, 140, 141
Road gravity, 93, 98
Road influenced growth, 93
Run-off, 22, 158, 141

S

Satellite, 52, 141
    ASTER, 115, 116, 118, 119, 122
    communication, 3
    data, 63, 141, 160
    Envisat, 116
    GOES, 115–117
    HCMM, 115
    image, 53
    Landsat, 5, 43, 47, 51–53, 64, 90, 96,
        115–119, 141
    meteorological, 114
    Meteosat, 8, 116
    MetOP, 116, 118
    MODIS, 115–117, 122
    Sentinel, 5
    SEVIRI, 116
    sun-synchronous, 118
    TIROS, 115
Scattering, 113
Scenario file, 95, 96
Secondary data, 52
Sediment trapping, 156
Sendai framework, 134, 146
Sensitivity, 133, 136
Sensors, 5, 46, 47, 52, 141
    AATSR, 116, 122
    AHS, 115
    airborne, 115, 119, 122
    Aqua, 116, 117
    ATLAS, 115
    AVHRR, 115, 116, 118
    calibration, 119
    Enhanced thematic mapper plus,
        115, 117
    hyperspectral, 117

space-borne, 2, 6, 52, 107
Terra, 116, 117
Thematic mapper, 115, 117, 118
thermal, 110, 115–119, 121
Sequester carbon
carbon sinks, 21
equilibrium, 21, 22
Shannon's entropy, 49, 63
Site suitability map, 151, 164
SLEUTH, 10, 86, 92–102, 151
SLEUTH-3r, 95
Slope resistance, 93, 98
Smart cities, 15
Spatial Decision Support System (SDSS), 149–168
Spatial, 108, 114–119, 121
analysis, 63, 136
changes, 2
coordinates, 3
data, 2, 5, 8, 47, 63, 141, 152
database, 153
features, 3
information, 2
metrics, 61–77, 92
pattern, 4
resolution, 5, 40, 47, 48, 101, 158
Special Economic Zones (SEZ), 154
Spectral
heterogeneity, 2
radiance, 112
reflectance curve, 49
resolution, 40, 47
Spontaneous growth, 93
Sprawl, 40, 74, 76, 168
Stephan-Boltzmann law, 109
Stubble burning, 28, 34
Supervised classification, 66
Supervised classifier, 7
Support vector machine, 43
Surface temperature, 45, 46
Sustainability, 127, 133
Sustainable, 23, 25, 122
development, 2, 98
management, 2, 161
planning, 133
Sustainable Development Goals (SDGs), 127–130, 144
Swath, 5, 48

T

Technology, 25, 31
Temperature
ambient air, 22, 119
at-satellite brightness, 119, 120
global mean, 22
in situ, 117
land surface, 107–109, 118, 121
Temporal
changes, 2, 5
pattern, 4
Thematic Mapper, 47
Transition
areas, 89
condition, 164
function, 87
probability matrix, 89
rules, 90–92, 151
Transmissivity, 114
True Colour Composite (TCC), 5, 6

U

United Nations Framework Convention on Climate Change (UNFCCC), 23, 24, 26
Unsupervised classification, 66
Unsupervised classifier, 7
Urban Growth Model (UGM), 83, 87, 92, 93
Urban Heat Island (UHI), 108, 109, 117–119
Urban
agglomerations, 64
centres, 1
climatology, 117
cover change, 42
environment, 2
expansion, 42, 90
floods, 130, 134, 135–136
growth, 13, 49, 76, 82, 86, 95, 96, 160, 168
heat islands, 45
model, 83, 85, 86
modelling, 77, 85, 91
morphology, 136
planning, 3
poverty, 128
sprawl, 63, 68, 102, 108
structure, 108

Urbanisation, 19, 35, 45, 46, 92, 118, 128, 134–137, 149–150, 159

V

Vector data model, 3, 4
Vegetation mapping, 44
Vegetation, 2, 48, 52, 107, 110, 111, 119, 162
Viewing angle, 110
Von Neumann, 87, 88, 90, 92
Vulnerability, 127–129, 133–134, 139–140

W

Water and food security, 22

Wavelength, 5
Weather
    extremes, 22
    forecasting, 115
Web Feature Service (WFS), 153
Web Map Service (WMS), 153
Web server, 159
Wein's displacement law, 112
Wetland ecosystem, 54, 156
Wind gust, 108

Z

Zenith angle, 120, 121
Zonal analysis, 69, 76